Monitoring the Comprehensive Nuclear-Test-Ban Treaty: Hydroacoustics

Edited by
Catherine deGroot-Hedlin
John Orcutt

2001

Springer Basel AG

Reprint from Pure and Applied Geophysics
(PAGEOPH), Volume 158 (2001), No. 3

Editors:

Catherine deGroot-Hedlin
Scripps Institution of Oceanography
La Jolla
CA 92093-0225
USA
e-mail: cdh@eos.ucsd.edu

John Orcutt
Scripps Institution of Oceanography
La Jolla
CA 92093-0225
USA
e-mail: jorcutt@igpp.ucsd.edu

A CIP catalogue record for this book is available from the Library of Congress,
Washington D.C., USA

Deutsche Bibliothek Cataloging-in-Publication Data

Monitoring the comprehensive nuclear test ban treaty. - Basel ; Boston ; Berlin : Birkhäuser
(Pageoph topical volumes)

Hydroacoustics/ed. by Catherine deGroot-Hedlin; John Orcutt. - 2001
ISBN 978-3-7643-6538-7   ISBN 978-3-0348-8270-5 (eBook)
DOI 10.1007/978-3-0348-8270-5

ISBN 978-3-7643-6538-7

© 2001 Springer Basel AG
Originally published by Birkhäuser Verlag in 2001

Printed on acid-free paper produced from chlorine-free pulp

9 8 7 6 5 4 3 2 1

Contents

Pure appl. geophys. 158 (2001) 421–422
0033–4533/01/030421–02 $ 1.50 + 0.20/0

❙ Pure and Applied Geophysics

Monitoring the Comprehensive Nuclear-Test-Ban Treaty

Preface

The first nuclear bomb was detonated in 1945, thus ushering in the nuclear age. A few political leaders quickly saw a need to limit nuclear weapons through international cooperation and the first proposals to do so were made later in that same year. The issue of nuclear testing, however, was not formally addressed until 1958 when the United States, the United Kingdom, and the Soviet Union, initiated talks intended to establish a total ban on that testing (a Comprehensive Test-Ban Treaty or CTBT). Those talks ended unsuccessfully, ostensibly because the participants could not agree on the issue of on-site verification.

Less comprehensive treaties did, however, place some constraints on nuclear testing. The United States, the United Kingdom, and the Soviet Union, in 1963, negotiated the Limited Test-Ban Treaty (LTBT) which prohibited nuclear explosions in the atmosphere, outer space and under water. The Threshold Test-Ban Treaty (TTBT), signed by the United States and the Soviet Union in 1974, limited the size, or yield, of explosions permitted in nuclear tests to 150 kilotons.

Seismological observations played an important role in monitoring compliance with those treaties. Many of the world's seismologists set aside other research projects and contributed to that effort. They devised new techniques and made important discoveries about the Earth's properties that affect our ability to detect nuclear events, to determine their yield, and to distinguish them from earthquakes. Seismologists are rightfully proud of their success in developing methods for monitoring compliance with the LTBT and TTBT.

Although seismologists have also worked for many years on research related to CTBT monitoring, events of recent years have caused them to redouble their efforts in that area. Between 1992 and 1996 Russia, France and the United States all placed moratoria on their nuclear testing, though France did carry out a few tests at the end of that period. In addition, the United States decided to use means other than testing to ensure the safety and reliability of its nuclear arsenal, and all three countries, as well as the United Kingdom, agreed to continue moratoria as long as no other country tested. Those developments, as well as diplomatic efforts by many nations, led to the renewal of multilateral talks on a CTBT that began in January 1994. The talks led to the Comprehensive Nuclear-Test-Ban Treaty. It was adopted by the

United Nations General Assembly on 10 September, 1996, and, as of late December 2000, has been signed by 160 nations. Entry of the treaty into force, however, is still uncertain since it requires ratification by all 44 nations that have some nuclear capability and, as of late December 2000, only 30 of those nations have done so.

Although entry of the CTBT into force is still uncertain, seismologists and scientists in related fields, such as radionuclides, have proceeded with new research on issues relevant to monitoring compliance with it. Results of that research may be used by the International Monitoring System, headquartered in Vienna, and by several national centers and individual institutions to monitor compliance with the CTBT. New issues associated with CTBT monitoring in the 21st century have presented scientists with many new challenges. They must now be able to effectively monitor compliance by several countries that have not previously been nuclear powers. Effective monitoring requires that we be able to detect and locate much smaller nuclear events than ever before and to distinguish them from small earthquakes and other types of explosions. We must have those capabilities in regions that are seismically active and geologically complex, and where seismic waves might not propagate efficiently.

Major research issues that have emerged for monitoring a CTBT are the precise location of events, and discrimination between nuclear explosions, earthquakes, and chemical explosions, even when those events are relatively small. These issues further require that we understand how seismic waves propagate in the solid Earth, the oceans and atmosphere, especially in regions that are structurally complex, where waves undergo scattering and, perhaps, a high degree of absorption. In addition, we must understand how processes occurring at the sources of explosions and earthquakes manifest themselves in recordings of ground motion.

Monitoring a CTBT has required, and will continue to require, the best efforts of some of the world's best seismologists and other scientists. They, with few exceptions, believe that methods and facilities that are currently in place will provide an effective means for monitoring a CTBT. Moreover, they expect that continuing improvements in those methods and facilities will make verification even more effective in the future. This topical series on several aspects of CTBT monitoring is intended to inform readers of the breadth of the CTBT research program, and of the significant progress that has been made toward effectively monitoring compliance with the CTBT.

The following set of papers, edited by Drs. Catherine deGroot-Hedlin and John Orcutt presents research results on hydroacoustic methods used to monitor a CTBT. It is the second of eight topics addressed by this important series on *Monitoring the Comprehensive Nuclear-Test-Ban Treaty*. The first Topic was Source Location and topics to appear in later issues are Regional Wave Propagation and Crustal Structure, Surface Waves, Source Processes and Explosion Yield Estimation, Infrasound, Source Discrimination, and Data Processing.

Brian J. Mitchell
Saint Louis University
Series Editor

Pure appl. geophys. 158 (2001) 423–424
0033–4533/01/030423–02 $ 1.50 + 0.20/0

| Pure and Applied Geophysics

Introduction

One of the challenges in designing a worldwide monitoring system for verification of compliance with the Comprehensive Nuclear-Test-Ban Treaty (CTBT) is that seismic networks can provide only sparse coverage of the oceans, which encompass more than 70% of earth's surface. Fortunately, acoustic energy created by underwater explosions can be trapped in the SOFAR (SOund Fixing And Ranging) channel and propagate to great distances with little attenuation, thus, relatively few stations are required for effective monitoring. The hydroacoustic component of the International Monitoring System (IMS), currently being installed for use in verifying compliance with the CTBT, will consist of six hydrophone stations and five supplemental T-phase stations located on ocean islands. This special volume brings together a number of studies addressing technical issues in detecting underwater explosions, and discriminating them from naturally occurring events.

The paper by Hanson *et al.*, describes the standard processing of hydroacoustic data at the Prototype International Data Center (PIDC) and specific studies conducted to validate results. In automatic processing, hydroacoustic signals are detected, classified and used to form events or associated to events formed by the other technologies. Nearly three years of processing has led to an extensive database that will be very useful in understanding basic scientific questions about seismic/acoustic coupling as well as improving our ability to effectively monitor a comprehensive nuclear-test-ban treaty.

The coupling between the underwater acoustic and the land seismic wavefields is the central issue in evaluating the performance of the T-phase stations in the hydroacoustic component of the International Monitoring System (IMS). This topic is addressed from a variety of perspectives in this volume; in terms of rays (Okal), acoustic modes (D'Spain *et al.*), a combined ray-mode approach (de Groot-Hedlin and Orcutt), and combined normal mode-finite difference modeling (Stevens *et al.*). Okal presents a study of T waves from nuclear tests conducted in Polynesia, recorded at the Hawaii Volcano Observatory network. Individual wave packets are interpreted in terms of conversion to either P or S energy, based on arrival times and spectral characteristics. In the paper by D'Spain *et al.*, data from a 3-km aperture vertical hydrophone array are used to estimate the normal mode components of naturally occurring T waves. As concluded in the paper, not only does the land/water coupling process cause a well-known decrease in signal amplitude, particularly at higher frequencies, but also the relative timing between the modes is altered by the coupling process, resulting in loss of information about the location of an in-water source. In de Groot-Hedlin and Orcutt, T-phase data generated by shallow earthquakes near Hawaii and by the French nuclear tests in Polynesia, are examined on the Berkeley seismic network records. The crustal velocities, slope geometry, and event-receiver azimuths are all shown to be controlling factors in acoustic to seismic conversion. Stevens *et al.* model T-phase signals with a composite techniques coupling normal mode propagation in the ocean to an elastic finite difference code to compute propagation on land. The computations reproduce characteristics of T phases recorded along the coast of California; namely, the spectral degradation of the seismic T phase, and that T phases propagate primarily as surface waves near the conversion point, but propagate mainly as P waves further inland.

Talandier and Okal present a discriminant separating earthquake and explosions based on the amplitude and duration of recorded T phases. The proposed discriminant is shown to be accurate for $M > 3$ and explosions greater than 80 kg, but is predicted to fail for smaller events.

Finally, the paper by Eneva *et al.* reports results from the analysis of unique Russian hydroacoustic data recorded at various depths and distances from charges detonated at various depths. These data are used to constrain the modeling of the hydroacoustic source as a function of explosion depth, so that the effects of detonating charges on or near the water surface are better understood. They show that the observations, including pressure measurements from 100-kg TNT explosions in a reservoir and a 1957 underwater nuclear explosion, match results predicted from a shockwave modeling code.

Catherine deGroot-Hedlin
John Orcutt
Scripps Institution of
Oceanography
La Jolla, CA 92093-0225
USA

Pure appl. geophys. 158 (2001) 425–456
0033–4533/01/030425–32 $ 1.50 + 0.20/0

© Birkhäuser Verlag, Basel, 2001

⎮Pure and Applied Geophysics

Operational Processing of Hydroacoustics at the Prototype International Data Center

JEFFREY HANSON,[1] RONAN LE BRAS,[1] PAUL DYSART,[2] DOUGLAS BRUMBAUGH,[1] ANNA GAULT,[3] and JERRY GUERN[1]

Abstract — The Prototype International Data Center (PIDC) has designed and implemented a system to process data from the International Monitoring System's hydroacoustic network. The automatic system detects and measures various signal characteristics that are then used to classify the signal into one of three categories. The detected signals are combined with the seismic and infrasonic detections to automatically form event hypotheses. The automatic results are reviewed by human analysts to form the Reviewed Event Bulletin (REB). Continuous processing of hydroacoustic data has been in place since May 1997 and during that time a large database of hydroacoustic signals has been accumulated. For a two-year period, the REB contains 13,582 *T* phases that are associated to 8,437 events. This is roughly 25% of REB events after taking station downtime into account. Predicted travel times used in locations are based on the arrival time of the peak amplitude mode calculated from a normal mode propagation model. Global sound velocity and bathymetry databases are used to obtain reliable 2-D, seasonally dependent, travel-time tables for each hydroacoustic station in the PIDC. A limited number of ground-truth observations indicate that the predicted travel times are good to within 5 seconds for paths extending to over 7,000 km – corresponding to a relative error of less than 0.1%. The ground truth indicates that the random errors in measuring arrival times for impulsive signals are between 1 and 6 seconds. This paper describes and evaluates the automatic hydroacoustic processing compared to the analyst reviewed results. In addition, special studies help characterize the overall performance of the hydroacoustic network.

Key words: CTBT verification, underwater acoustics, International Data Center.

1. Introduction

1.1. The Prototype International Data Center

The Prototype International Data Center (PIDC) is being developed to demonstrate methods to monitor compliance with the Comprehensive Nuclear-Test-Ban Treaty (CTBT), using the four technologies specified in the treaty (seismic,

---

[1] SAIC – Monitoring Systems Operations, 10260 Campus Point Dr., San Diego, CA 92121, USA.
[2] SAIC – Ocean Sciences Group, 1710 Goodridge Dr., McLean, VA 22102, USA.
[3] SAIC – Center for Monitoring Research, 1300 N. 17th Street, Suite 1450, Arlington, VA 22209, USA.
Corresponding author: Jeffrey Hanson, E-mail: jhanson@gso.saic.com

hydroacoustic, infrasonic and radionuclide). The PIDC began as the Group of Scientific Experts Third Technical Test (GSETT-3) which was an experimental global seismic monitoring system (RINGDAL, 1996). Since then it has expanded to include the other technologies, although seismic processing remains the most developed and mature of the four.

The products of the PIDC are a series of global event lists and bulletins which contain hypothesized event locations and their associated signals. There are three standard event lists produced by the automatic system (SEL1, SEL2 and SEL3). The SEL1 is produced very soon after the time of occurrence of the events. An event will be listed in that bulletin two hours after its origin time. This earliest of automatic bulletins is produced using only the primary global network of seismic and hydroacoustic stations. The events in the SEL1 are used to predict arrival times of seismic phases and request relevant waveform segments from the auxiliary stations. The detections from these additional data are used with the primary stations' detections to form the SEL2. The SEL3, which is produced approximately 12 hours after real time, is nearly identical to the SEL2 but allows for late arriving data. The SEL3 is used as the starting point for interactive processing. Analysts review each event and all waveform data, refining arrival times, adding missed detections and removing incorrect associations. Events are then relocated with the refined data. Events which meet a set of established criteria are included in the Reviewed Event Bulletin (REB). Further bulletins are produced which contain parameters to aid in event characterization. In this paper we will use the SEL3 as the automatic bulletin and the REB as the reviewed bulletin. For further information on the products of the PIDC visit their web page (www.pidc.org).

1.2. The Hydroacoustic Network

The International Monitoring System (IMS), which is responsible for maintaining the international network of instruments, is installing a hydroacoustic network consisting of 11 stations (Fig. 1). The network consists of two station types: hydrophone stations and "*T*-phase" stations. *T*-phase stations are seismic stations intended to record signals that mainly propagate through the oceans. Both hydrophone and *T*-phase stations are considered primary stations within the IMS.

Hydrophone stations will consist of hydrophones located offshore from mid-ocean islands. In most cases the hydrophones will be connected to the island with fiber optic cables extending approximately 100 km on two sides of the island to minimize local blockage. The hydrophones will float at a depth corresponding to the local SOFAR channel axis. The digitizer is located at the instrument end of the cable in order to maximize the recordable dynamic range of the hydrophone. The new station specifications require that there be three hydrophones at the end of each cable, two of which are nominally designated as backup instruments. However, it is anticipated that the three hydrophones will be deployed in a triangular pattern with a

Planned IMS Hydroacoustic Network

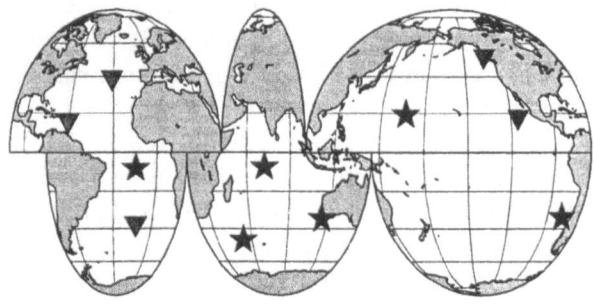

PIDC Hydroacoustic Network

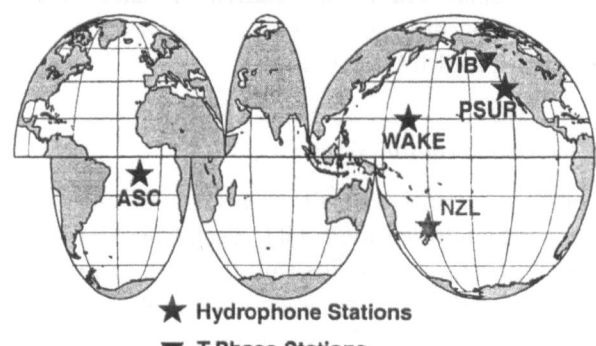

★ Hydrophone Stations

▼ T-Phase Stations

Figure 1
The planned IMS network has 6 hydrophone stations and 5 T-phase stations. The hydrophone stations, which are more sensitive than T-phase stations, are mainly located in the southern oceans to compensate for the seismic network's lack of coverage. A hydrophone station will generally consist of two or more widely spaced instruments to reduce blockage and provide azimuthal constraints. The current network used at the PIDC has 3 hydrophone stations and 1 T-phase station. A fourth hydrophone station near New Zealand was used temporarily. The Wake and Ascension stations consist of two and three hydrophones, respectively.

spacing of 1 to 2 km in order to provide bearing estimation capabilities (e.g., HANSON and GIVEN, 1998). Five of the six hydrophone stations will be located in the southern oceans where seismic network coverage is relatively sparse.

T-phase stations will consist of one or more seismometers located near-shore, typically on mid-ocean islands. It is expected that they will record signals that have primarily traveled through the oceans, converting into elastic waves at the ocean island interface. The performance of T-phase stations is not well known, particularly at high frequencies (> 10 Hz) (e.g., TALANDIER and OKAL, 1998; PISERCHIA et al., 1998; STEVENS et al., 1998; HANSON et al., 1997). They were conceived as a cost-saving measure because hydroacoustic stations are very expensive, largely due to the long cable extending offshore. The T-phase stations will be mainly located in the Northern Hemisphere where the seismic network coverage is relatively dense.

The current network in use at the PIDC is considerably smaller than the proposed IMS network (Fig. 1). It currently consists of 3 hydrophone stations (Wake Island, Ascension Island and Point Sur, California) and 1 T-phase station (Victoria Island, British Columbia). A fourth hydrophone station off New Zealand's coast was used on a temporary basis for more than a year. The Wake Island station has two hydrophones (WK30 and WK31) separated by 240 km, and the Ascension Island station has three hydrophones (ASC23, ASC24 and ASC26) with station spacings ranging between 3 and 100 km. The Point Sur station has been used at the PIDC in lieu of other data and only has 1 hydrophone (PSUR). It will *not* be part of the IMS network. The New Zealand station consisted of two hydrophones (NZL01 and NZL06) spaced 0.6 km apart. The T-phase station on Victoria Island (VIB) consists of a single vertical component seismometer. The stations at Wake, Ascension and Victoria Islands are proposed IMS stations.

1.3. Objective of the Hydroacoustic System

The hydroacoustic processing at the PIDC has been developed to demonstrate an effective approach for monitoring the CTBT throughout the world's oceans. This includes integrating hydroacoustic results with the other technologies. The primary goal is to detect, identify and locate large underwater explosions. This is to be done in conjunction with the seismic and infrasonic networks (BRATT, 1996; BROWN et al., 1998). In addition, the PIDC associates hydroacoustic signals (T phases) from earthquakes which have been located using the seismic network and can be useful in event characterization. The bulletin resulting from this processing is dominated by earthquake T phases because they are quite common, as opposed to in-water explosions which are rare occurrences. The hydroacoustic network may also be used to help monitor for low atmospheric nuclear explosions which are thought to weakly couple energy into the SOFAR channel (CLARKE et al., 1997; BACHE et al., 1980). For this scenario the hydroacoustic network would likely need to be used in conjunction with the infrasound and radionuclide networks because the expected low-amplitude, low-frequency signal is difficult to distinguish from small T phases or other "noise" phases.

The ocean waveguide is such an efficient transmitter of sound that a large in-water explosion will produce a large signal at any hydrophone that has an unblocked path, and even hydrophones that do not have a direct path to the event are likely to see a signal because of multi-path effects (ANGELL et al., 1998). One method to monitor the oceans is to only examine large signals and ignore all others. This approach simplifies the hydroacoustic processing procedure, however it misses a wealth of information available for characterizing events near the ocean boundaries. Most natural events occur near oceans, and many of these produce hydroacoustic signals (currently 25% of the REB events). Hydroacoustic signals from earthquakes tend to be quite different in duration than those from underground explosions (MILNE, 1959), thus the hydroacoustic signals provide a potentially powerful discriminant between natural and man-made events. Before this discriminant can be used, more theoretical research needs to be conducted to constrain the range of waveform characteristics that a near-shore, underwater explosion can create.

1.4. T, H and N Phases

At the PIDC, hydroacoustic signals are classified into three categories: T, H and N. The traditional phase designation in seismology for a phase whose path includes propagation through the ocean is T for tertiary because the signal arrives after the primary (P) and secondary (S) phases (LAY and WALLACE, 1995). The commonly observed T phase is a long duration (minutes), band-limited (2–40 Hz) wavetrain and is commonly generated by earthquakes near ocean boundaries. The T-phase characteristics do not differ dramatically between seismic and hydrophone recordings, although the higher frequencies can be more attenuated on land. Earthquake generated hydroacoustic signals have been used to locate small events on mid-ocean ridges (FOX et al., 1995), but the complicated coupling mechanism (DE GROOT et al., 1998) producing the T phase generally prevents their use in hypocenter estimation on a global scale.

An in-water impulsive event, such as an explosion, generates a substantially different signal than the earthquake generated signal. It is short in duration (< 45 seconds), and broad in frequency content (2–100 + Hz) (e.g., LAVERGNE, 1970; CHAPMAN, 1985; URICK, 1983). The hydroacoustic signals from in-water explosions are very useful in locating events because the ocean sound speed is well known and the impulsive source function is small in both duration and spatial extent. Due to the significantly different characteristics between these signal types, another phase designator is needed. Signals that appear to be generated by land-based events are still called T phases, whereas signals that appear to be from in-water impulsive sources are designated as H phases.

It was decided not to attempt designating signals based on the mode of propagation for each portion of the path as done in seismology. For example, an earthquake generated signal which propagates through the ocean and is then recorded on a T-phase station might have been called "PTP". The problem with this approach is that the mode of propagation on land is not known and probably includes compressional, shear and surface type waves (PISERCHIA et al., 1998; OKAL and TALANDIER, 1997; STEVENS et al., 1998). It would be similar to separating the regional seismic Lg phase into its various components. The main reason for differentiating between T and H phases at the event formation stage is to determine whether or not to use the signal in locating events. The fact that the two phases indicate very different sources is important, but here we only evaluate the system's event/signal association and location abilities and not its event characterization ability.

Because of the efficiency of sound propagation in the oceans, hydrophones record many transient signals, most of which are not of interest to the monitoring community. These can include various shipping noise and biologically generated sounds such as those from marine mammals. The automatic system attempts to classify these signals as "noise" or N phases based on waveform characteristics measured by the automatic system, however distinguishing between N phases and small H phases is difficult.

2. Hydroacoustic Processing Overview

2.1. Processing Description

The hydroacoustic processing at the PIDC follows very similar steps to those used in seismic processing (Fig. 2). An automatic system processes waveform data and produces a series of standard event lists (LANEY *et al.*, 1996). These are used as a starting point for interactive processing in which analysts review all of the events and detections made in the automatic system. The tasks in the automatic processing include: detection of signals, extraction of signal characteristics, preliminary phase identification, association of phases to events, and, in certain cases, a preliminary location based on arrival times.

2.2. Detection and Feature Extraction

Detecting signals in hydroacoustic data is accomplished using a short-term average to long-term average (STA/LTA) type detector. The time windows used for

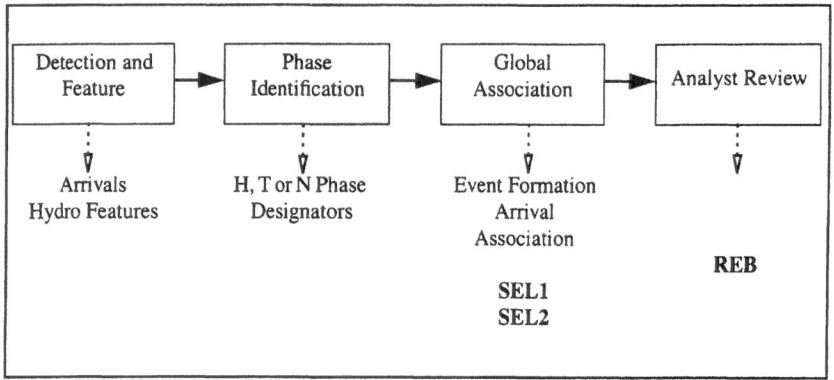

Figure 2
Basic flow chart of hydroacoustic processing and the general results from each step.

the short-term and long-term averages are 10 and 150 seconds (Fig. 3), which are longer than those used for seismic data – typically 1 and 60 seconds (WILLEMANN, 1998). This avoids overwhelming the system from local signals created by, among other things, biological sources which are typically short in duration. The detector is used across several frequency bands and the threshold level is set on a station-to-station basis.

A set of signal features are estimated and stored for each detection (Table 1). They are used in later processing to identify phase types and can be useful in event

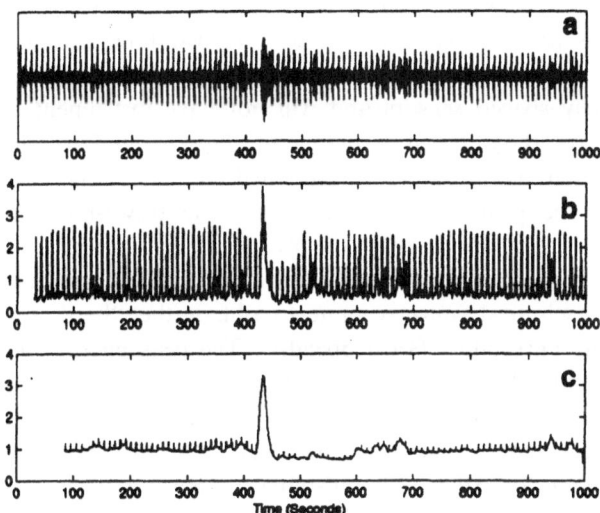

Figure 3

Hydroacoustic signal detection in the presence of interfering signals. **a**) The raw waveform data. The signal of interest (at 420 seconds) is visible, however a series of clicks (most likely from whales) makes automatic detection difficult. **b**) The STA/LTA average using windows of 1 and 60 seconds, respectively (typical windows for seismic processing). **c**) Same as **b**, but for window lengths used in hydroacoustic processing (10 and 150 seconds).

Table 1

Major features estimated for each detection

Features	Description
Onset and termination times	The estimated beginning and the end of the signal
Arrival time	Three arrival time estimates:
	peak energy
	mean energy
	probability weighted
Signal amplitude measures	Peak energy and total energy
Cepstral measurements	Delay time and level of maximum peak in cepstrum
Energy time distribution measurements	Time spread, energy skewness, etc.

characterization. Each feature is measured over a set of frequency bands in order to observe variations with frequency (typical frequency bands are 2–4, 3–6, 4–8, 6–12, 8–16, 16–32, 32–64 and 2–80 Hz). The first features to be estimated are the onset and termination times. These define the waveform segment used when estimating the remaining features.

The arrival time of a hydroacoustic phase is not as well defined as it is for seismic body waves. The acoustic waves travel through the oceanic waveguide, and hence are composed of multiple ray paths or modes (JENSEN *et al.*, 1994). This makes the hydroacoustic phase more akin to seismic surface waves than to body phases. However, the sound velocity structure of the oceans is very well known, and highly precise locations can be made if a reliable arrival time can be estimated and associated to the correct mode of propagation. The IMS hydroacoustic network consists of widely spaced sensors and will not have instrument arrays. Without hydrophone arrays, the different modes cannot be separated, and hence the arrival times of the different modes cannot be estimated separately. Instead, we estimate the arrival time of the peak energy, after which the travel-time models used for association and location are based on the predicted highest energy mode (see Section 2.5). The peak arrival time is estimated using a probabilistic approach that takes the uncertainty in each peak into account to obtain the "best" estimate and the associated timing uncertainty (see Appendix). The peak energy from an explosive source can be accurately modeled, but the peak energy of a *T* phase generated by an earthquake is not well understood. The long duration of *T* phases and the likely distributed source region at which they are generated makes the peak arrival time less meaningful than it is for explosive point sources. A more stable measure – at least for distances of tens to hundreds of kilometers – is the mean energy arrival time. Although this arrival time is more stable between sensors, how it relates to the event's location is not well understood. The mean energy arrival time is computed and stored, but is not used as the final arrival time.

A variety of other measures are made to characterize the signal. Peak amplitude and total energy indicate the size of the signal. Time spread, skewness and kurtosis are moment statistics quantifying the energy distribution with time. Cepstral measurements that quantify the scalloping in the signal's power spectrum are also made (see Section 4.2) and are useful in characterizing underwater explosions (GITTERMAN *et al.*, 1998). All these metrics can be useful in differentiating among *T*-, *H*- and *N*-type phases and can also be used to constrain certain source parameters.

2.3. Phase Classification

Station processing of hydroacoustic signals includes the identification of arrivals at hydroacoustic stations using either a trained artificial neural network (NNET) applied in two stages, or a single set of default rules. There are three definitive phases used to identify hydroacoustic arrivals (noise, *T* phase, and *H* phase). In the case of noise and T-phase arrivals, phase identification is based on features extracted from many data examples. Noise arrivals from sources such as marine mammals and shipping activity, and T phases associated with events recorded by the seismic network, are routinely recorded by the hydroacoustic network. These data provide ample NNET training data. However, due to the scarcity of in-water explosions, the identification of H phases relies mostly on features extracted from broadband time-series simulations. These are generated using the normal-mode model KRAKEN modified by an interpolation scheme that makes broadband time series simulation a practical alternative (PORTER and REISS, 1985; LANEY, 1994).

A so-called three-layer back-propagation neural network (RUMELHART and MCCLELLAND, 1986) is used for phase identification at hydroacoustic stations. The three-layer network consists of an input layer, a hidden layer, and an output layer. Each layer is made of a set of "nodes." The input layer simply has one node per input feature, and the output layer has one node per phase type. The number of nodes in the hidden layer is determined during training through a selection process that attempts to optimize the classification results. Each node in the hidden layer takes a weighted average of the input values and transforms this weighted average using an activation function (e.g., BISHOP, 1995). The weights in the previous sum are known as connection weights, and the activation function is a smooth function that approximates a step. The output of each node is used as the input in the next layer. Neural networks can have multiple hidden layers, however the use of one hidden layer is justified by its well documented mapping capabilities (e.g., HORNIK, 1991; KURKOVA, 1991).

The neural network is trained using an input data set of known phases. Training the neural network is essentially a process of adjusting the connection weights to optimize its classification properties. The method used in determining the weights is a type of gradient-descent algorithm that is subject to many of the same constraints as other conventional minimization techniques. Like other classifiers, the data (training set) and the number of coefficients in the objective function (connection weights) must be selected and balanced to ensure a solution that incorporates a general relationship between the data and the known class to which it belongs. Otherwise a situation known as overtraining may occur which produces a neural network that works well for the training set, but performs sub-optimally on a more general data set.

To select parameters appropriate for NNET training, the raw time and amplitude measures extracted from the hydroacoustic signals (Table 1) are first used to derive a set of parameters that are insensitive to event magnitude and absolute times. This is done to avoid problems in pattern recognition arising from parameter magnitude and scaling. The parameters are then prioritized, based on how well they distinguish between the different phase types, using a measure known as the Mahalanobis distance (DUDA and HART, 1973) and a measure of the covariance between each input parameter. Parameters that separate the phase types the most and are independent of the other parameters are given the highest priority. These pairwise and single parameter linear measures are effective guides to parameter selection and network design.

The phase classifier used in the system is actually two neural networks in serial. The need for a two-stage approach is indicated by the similarity in waveform character between H and N phases. Therefore the first stage NNET attempts to separate H and N phases from T-phase arrivals. In the second stage a different set of parameters is used to separate H and N phases. A list of the most useful parameters for arrivals at PSUR and WK30 is given in Table 2. The SIGNAL_IN_BAND parameter is determined by an SNR threshold of 4.

The default rules, used to identify hydroacoustic phases at stations where a set of NNET weights is not available, are largely based on the visual observations of many signal types. The rules currently in place at PSUR and WK30 are simple thresholds on the duration, time spread, and signal_in_band parameters in the 3–6 Hz and 32–64 Hz bands. These default rules, based on the duration and frequency content of arrivals observed at PSUR and WK30, should be effective at other deep hydrophone sites, but were not intended for use at T-phase stations where waveform characteristics can vary greatly. The confident identification of explosions at T-phase stations represents a significant challenge to be addressed as more of these stations become operational.

2.4. Global Phase Association

Once detections have been made on the waveforms from hydroacoustic stations and an initial identification as earthquake-generated (T), explosion-generated (H) or noise (N) established, the next major processing step is to associate together extracted arrivals from different stations and thus form events automatically. That step is performed using a program called Global Association (GA).

Table 2

Neural network input parameters

STAGE 1: N/H-TARGET T-CLUTTER		STAGE 2: H-TARGET N-CLUTTER	
PARAMETER	BAND	PARAMETER	BAND
MEAN_DELAY (sec)	6–12 Hz	SIGNAL_IN_BAND	6–12 Hz
MEAN_DELAY (sec)	4–8 Hz	PEAK_DELAY-MEAN_DELAY	3–6 Hz
MEAN_DELAY (sec)	3–6 Hz	PEAK_DELAY-MEAN_DELAY	4–8 Hz
DURATION (sec)	6–12 Hz	SIGNAL_IN_BAND	2–4 Hz
SIGNAL_IN_BAND	32–64 Hz	SIGNAL_IN_BAND	8–16 Hz
DURATION (sec)	4–8 Hz	SIGNAL_IN_BAND	2–80 Hz
TIME_SPREAD (sec)	3–6 Hz	SIGNAL_IN_BAND	16–32 Hz

GA is a grid-based exhaustive search algorithm that was originally designed to process seismic detections and has since been upgraded and adapted to process hydroacoustic and infrasonic data (Fig. 4). A GA process is triggered at regular time intervals, usually every 20 minutes, and the newly detected set of arrivals in the latest 20 minute interval is scanned for arrivals from different stations that belong together in one event. The principal steps in forming events automatically within GA are the same independent of the technology. A number of adaptations have been added to take into account the particularities of each technology. In addition to summarizing the processing steps of the automatic association process, the following discussion outlines the adaptations made to the algorithm in order to handle hydroacoustic processing.

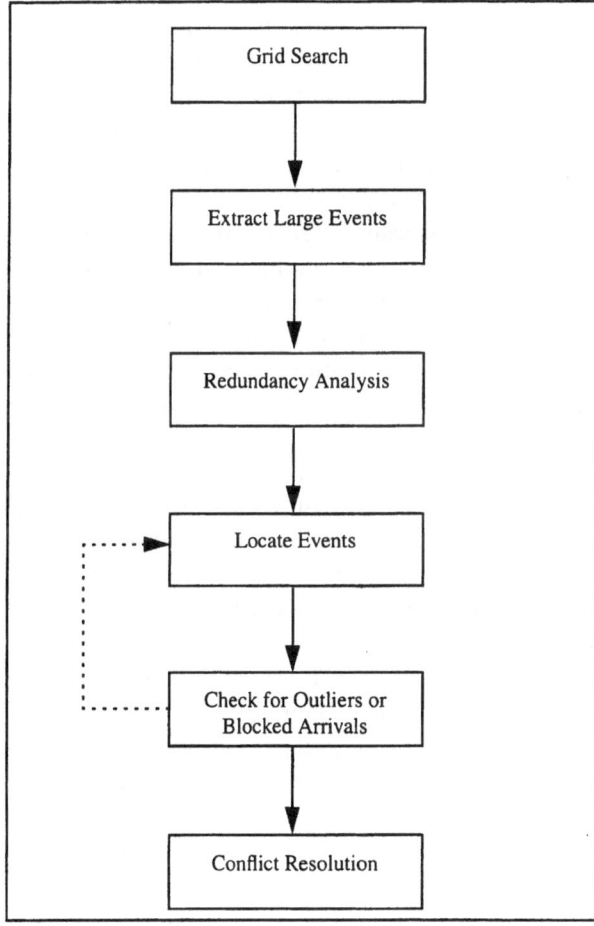

Figure 4
A flow diagram for the Global Association module. See text for the details in each step.

The first step is an exhaustive search where at each grid point either on the surface of the earth or at depth for areas with deep seismicity, the arrival set at stations in the network is searched for all possible events that may have happened in a circular region around the grid point. We refer to that circular region as the grid cell. Typically, in the global problem, the spacing between grid points is about 3 degrees and the radius of the grid cell is about 2.5 degrees. A pre-established network-specific knowledge base is built prior to the exhaustive search, where travel time and slowness characteristics for all phases expected to be seen at each station in the network are stored for each grid cell. For hydroacoustic processing, blockage is taken into account in the sense that the hydroacoustic phase due to a marine explosion is not stored within the knowledge base if all possible paths are blocked between a hydroacoustic station and points within the grid cell.

The mechanism for forming sets of associations at a grid point is based on the concept of a *driver* arrival and a search for arrivals at corroborating stations compatible with that *driver* arrival. A *driver* arrival is identified as an arrival at one of the stations closest to the grid cell (the number of stations scanned is adjustable; we usually use the five closest stations) that has an azimuth and slowness compatible with an event within the grid cell. In the prototype IDC implementation, the *driver* arrival may be an *H* phase at a hydroacoustic station. In a typical 20 minute interval, 500 to several thousands of association sets are produced at this stage.

The next step is to extract large events and their associated arrivals. The motivation is to make the processing more efficient and allow events with a multitude of defining phases to be formed, located, and their conflicts with other events resolved early in the automatic association process. This step mostly affects events with a large number of associated seismic arrivals and has little incidence on hydroacoustic processing.

A redundancy analysis is then performed that eliminates events whose association sets are entirely a subset of another event. Additional constraints are that the two events share a common *driver* arrival and that phase identifications for all common arrivals be the same in both events. This redundancy step typically reduces the total number of preliminary events by a factor of ten.

After redundancy analysis, an event location and outlier analysis step is performed on the preliminary events. The location is performed using an iterative travel-time table driven least-squares algorithm. The algorithm has the option to perform fixed depth or free depth locations. The initial location for the iterative process is the center of the cell where the preliminary event was first formed. The outlier analysis is performed immediately after location. It consists of examining the RMS of normalized residuals of the time and slowness components for the location on an arrival per arrival basis, for all defining arrivals (arrivals whose time and/or slowness are used to define the location). If the residuals exceed a *chi-square* test based criterion, the arrival with the worst residual is eliminated from the association set and the location re-evaluated with the reduced association set. The located event is repeatedly subjected to the outlier analysis until it is free of outlying arrivals. For hydroacoustic phases an additional check is performed after location to ensure that the path between event and station is clear. If a path is found to be blocked, the hydroacoustic association is removed and the event re-located.

The last step in automatic association is the conflict resolution step in which arrivals that may have previously been associated to multiple events are analyzed and a single association selected from the multiple possibilities. The algorithm used to perform this conflict resolution step uses a metric with both event-based and association-based components. The event-based metric is itself made up of several components, whose relative importance are parameterized and can be adjusted. The number of defining associations and the size of the error ellipse are usually weighted the most. The association-based component is a measure of the normalized residuals.

There is a greater likelihood for hydroacoustic arrivals to be associated to events with seismic and hydroacoustic phases (hybrid events) than there is for seismic arrivals. Because of this a specialized event-based metric is used for hydroacoustic associations. The metric is a weighted sum of the number qf hydroacoustic arrivals and non-hydroacoustic arrivals. If the desired effect is to group together the most hydroacoustic arrivals, the former is predominately weighted. After events are formed through the process described above, they can be augmented and improved using an event-based predictor for both defining phases and non-defining phases. A defining phase contributes to the hypocenter's calculation, whereas a non-defining phase is simply associated to an event. T phases are non-defining in the current system, and it is at this final stage that they are added to the events.

2.5. Travel-time Tables and Blockage Maps

Hydroacoustic processing uses either a default 1-D travel-time table (constant velocity 1485 m/s), or a table specific to a station and season to aid in event association and localization. Travel time (TT) and transmission loss (TL) are calculated for each station in the network. Blockage maps are derived from the TT/TL tables. Signals are considered blocked when the environmental databases indicate a landmass or the TL exceeds 225 dB. A two degree tolerance factor is added to the perimeter of the blockage map to allow for signals generated from in-land events. Currently, 2-D seasonal tables including weighted travel times, transmission loss, and model error estimates are available for all operational stations.

The TT/TL tables were developed using the Active Sonar Performance Model (ASPM) (BURGER et al., 1994) to predict TT and TL from each station to any point on a globe sampled every 30 nautical miles (nm) in range and 0.5 degrees in bearing. The propagation method within ASPM is ASTRAL, an adiabatic ray-mode model which averages TT and TL over oceanic convergence zones of approximately 30 nm. Predictions are based on the Navy standard 5 minute bathymetry database DBDB5, and the standard Navy sound speed profiles for summer, autumn, winter and spring (GDEM). All acoustic simulations assume a 10 m source depth and 10 Hz acoustic propagation frequency. The models are computed along geodesic paths radiating radially outward from the stations. There is no attempt to compensate for horizontal refraction which may occur along some of the high latitude paths (SHOCKLEY *et al.*, 1982). The calculation is terminated at the range along each bearing where the total transmission loss exceeds 225 dB. At this point the model errors become large and erratic due to the elimination of all but a few modes.

To be compatible with the measured arrival times, travel times and modeling errors are computed using a probability-weighting algorithm (see Appendix) applied to synthetic multimode wavetrains generated by the ASPM software. The resulting travel times and errors take into account the uncertainty associated with determining the peak amplitude mode within the modeled wavetrain. Analysis of the modeling results for all stations and seasons indicates that the modeling error for the 10 m source is of the same order as errors due to source depth uncertainty. Additional error terms have been included to account for modeling errors due to sound speed fluctuations over spatial and temporal scales. The total model error is

$$\sigma_{\text{total}} = \sqrt{\sigma_{pwt}^2 + \frac{4\Delta}{90°} + \frac{N^2 + P^2}{2}}$$

where σ_{pwt} is the uncertainty from the probability weighted time calculation, and Δ is distance in degrees. P and N are the differences in travel time between the current season and the previous and following seasons, respectively.

The quality of travel-time predictions for T phases associated with seismic events is generally not favorable due to uncertainties in the source/coupling location. At T-phase stations the problem is compounded by uncertainties in the propagation of signals at the ocean-island interface. This represents a challenge to the effective use of data recorded at island and near shore stations. However, analysis of the few impulsive data examples available (Section 4.1) indicates that ASPM modeled travel times are reliable to within a few seconds, with limitations and biases most likely imposed by the resolution of the sound speed databases.

2.6. Interactive Review

The results processed by the automatic system must undergo analyst review before migrating to the Reviewed Event Bulletin (REB). Although the overwhelming majority of hydroacoustic signals examined by the analysts are T phases associated with earthquake activity, the interactive system is designed for detailed analysis of both T phases and H phases. Analysts examine the waveforms and events built by the automatic system through an interactive display known as the Analyst Review Station (ARS). Analysts identify T phases by their duration and low peak frequency, while H-phase identifications are based on their impulsive nature and high frequency energy content. Table 3 provides a brief outline of characteristics used to distinguish between T and H phases. These are guidelines and not strict rules. For example, a large explosion can have a prolonged duration due to phases reflected off coastlines or other bathymetric features.

Certain parameters that are measured by the automatic system are displayed in ARS for the analyst to evaluate. These include the optimal filter band at which the highest total energy is recorded, as well as signal onset and termination times. When the analyst examines the waveforms and detections in ARS, they are asked to judge these parameters. They evaluate the filter band chosen by the automatic system, and make adjustments if they feel that another filter is more appropriate, based on visual assessment of signal-to-noise ratio. Analysts may also re-calculate signal parameters based on adjusted signal onset and termination times. Finally, the analyst evaluates the automatic arrival time estimate and will make adjustments if necessary.

After positively identifying the signal as a hydroacoustic phase, analysts must determine if it has been appropriately associated to the correct event. ARS has the capability to display hydroacoustic blockage maps that analysts can consult before saving an event. Finally, analysts must keep in mind that certain time residual restrictions exist for hydroacoustic phase association. T phases are allowed a time residual up to 120 seconds because of the uncertainty in T-phase coupling, and H phases must be within 60 seconds of the theoretical arrival time.

3. Results from the Global Data Set

3.1. The Reviewed Event Bulletin and the Standard Event List

The REB provides a large database of analyzed events from which to evaluate the automatic system (SEL). It is the largest global database available that associates T phases to events (Fig. 5), however two caveats must be kept in mind. 1. T phases are only verified by analysts if they are associated to events that meet the REB criteria. 2. There have been no large underwater explosions since the beginning of hydroacoustic processing. The SEL database tables contain an arrival for each detection made whether or not it is associated to an event, whereas the

Table 3

Signal characteristics

	H	T
Frequency Content (Hz)	1–100 +	1–40 (peak usually < 10 Hz)
Duration (sec)	< 45 sec	30–240 +
Shape	Impulsive, high SNR	Emergent
Spectrum	Full band, possible scalloping	Decaying over 20 Hz

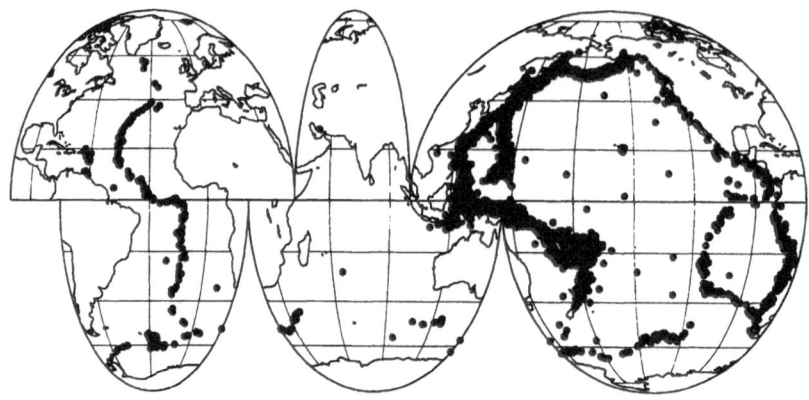

Figure 5

REB events with T phases between May, 1997 and May, 1999. There are 8,437 events in the two-year period that have an observed T phase at at least one hydrophone station. Most events are along the western edge of the Pacific. A few events appear questionable, however they comprise a very small percentage of the total ($<0.1\%$).

REB only contains arrivals associated to events. This means that many T phases in the SEL that are not in the REB are, in fact, real. In evaluating the automatic system relative to the REB we cannot determine the exact misclassification rate of T phases in the SEL that do not make it into the REB, however we can examine how much of the REB is produced by the automatic system. In addition, special studies have been conducted to measure the system's performance in classifying H phases which provides confidence that the system is able to distinguish between phase types.

There are general observations about hydroacoustic signals in the bulletins that can be made. Spanning two years 13,582 T phases have been associated to 8,437 events in the REB (Fig. 5). Considering station downtime, this roughly indicates that 25% of all REB events have an associated T phase. This percentage should increase by a few percent with the introduction of the new IMS hydrophone stations, particularly with the Indian Ocean stations. Even though the completed IMS hydroacoustic network will have significantly better coverage than the current network, the expected increase in REB associations is moderate because most near or below ocean events likely to produce T phases occur in the western Pacific, which is well covered by the Wake Island station. A handful of events in Figure 5 clearly have T phases incorrectly associated to them, and are likely due to human error. For example, the event in the Persian Gulf is unlikely to have created a T phase which could travel to any of the current stations. However, these events represent a very small fraction of the total ($<0.1\%$). Below, we discuss each station individually and then compare results to classify performance based on station characteristics. Tables 4 and 5 provide the overall statistics for each of the stations.

Table 4

Average number of detections

	PSUR	WAKE	ASC	NZL	VIB
T phases detected in SEL per day	14	28	11	4	11
N phases detected in SEL per day	9	52	43	94	34
H phases detected in SEL per day	1.5	1.0	0	6.3	0
Total phases detected in SEL per day	24	81	54	104	45
T phases associated in SEL per day	3.9	10	1	0.6	1.5
H phases associated in SEL per day	0.9	0.2	0	3	0
T phases associated in REB per day	4.6	11	1.5	0.3	0.8
Total # of detections since start of processing	13,246	59,458	16,790	32,863	25,854

Table 5

Performance of automatic processing

	PSUR	WAKE	ASC	NZL	VIB
T phases associated in REB per day	4.6	11	1.5	0.3	0.8
% detected in SEL	77	85	81	70	61
% of these detections correctly identified as *T*	90	89	83	32	78
% of correctly identified *T* phases associated to same event in SEL and REB	71	73	65	75	67

3.2. Wake Island

The Wake Island hydrophone station is part of the IMS hydroacoustic network. It has been in operation since the beginning of standard hydroacoustic processing at the PIDC. Initially data were received from one hydrophone (WK30), but in February 1998 data from a second hydrophone (WK31) became available. The hydrophones are part of the U.S. Air Force Missile Impact Locating System (MILS) that were installed in the 1950s and 1960s to aid in the location of missile splash downs (McCREERY and WALKER, 1988). The signals from the hydrophones are transmitted through underwater cables and digitized on the island (320 Hz sample rate). The hydrophones are located above seamounts, and float in the SOFAR channel ($\sim$800 meter depth). The main difference between the current instruments and the specifications for planned IMS hydrophone stations is the dynamic range of the instruments. The dynamic range is anticipated to increase from roughly 40 dB to 120 dB.

There are on average 10 *T* phases per day associated to REB events. This is the highest association rate in the current network and is due to the station's proximity to the convergent margins in the western Pacific (Fig. 6). The associations represent approximately 25% of the REB events that the blockage maps predict could be seen at Wake.

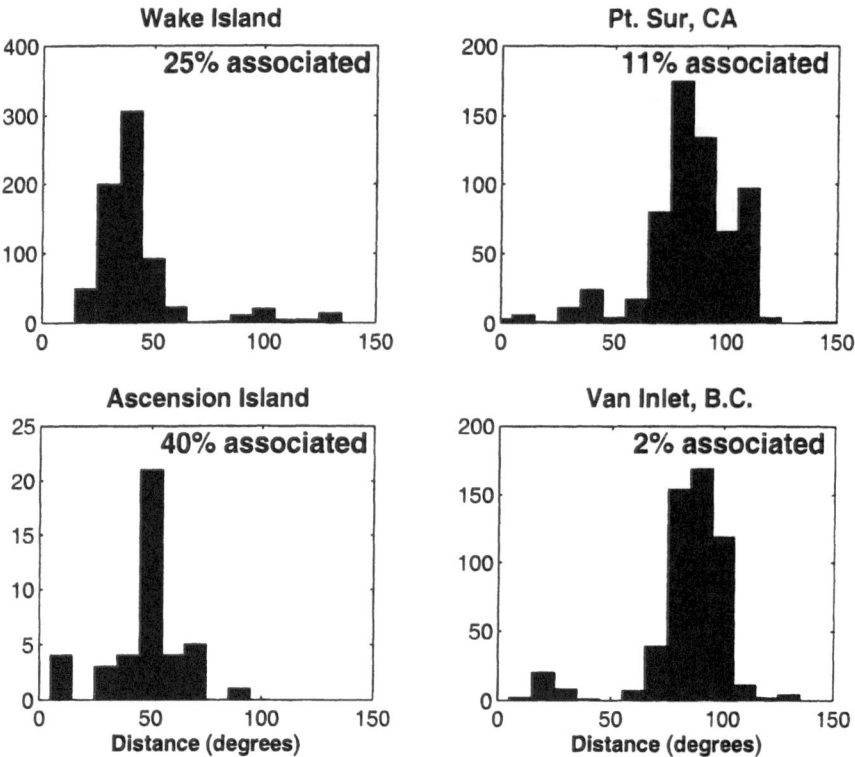

Figure 6

Distance histograms for one month of REB events with unblocked or nearly unblocked paths to each station. Most of these events do not have associated *T* phases (75% at Wake, 89% at Pt. Sur, 60% at Ascension and 98% at VIB). The difference in *T* phase association rate between Wake and Pt. Sur is likely due to the differences in the event distance distributions. This does not explain the difference between Pt. Sur and VIB which is due to the lower sensitivity of *T*-phase stations.

3.3. *Ascension Island*

The hydrophones at Ascension Island are also part of U.S. Air Force Missile Impact Locating System. Ascension is one of the proposed IMS hydroacoustic stations. The station is very similar to the Wake Island station, with the exception of the sampling rate (120 Hz compare to 320 Hz at Wake). The data received at the PIDC have been intermittent, but currently the PIDC is receiving continuous data from 3 hydrophones (ASC23, ASC24 and ASC26).

The Ascension Island hydrophones have, on average, 1.5 *T* phases associated to REB events per day. Because Ascension is located in the equatorial Atlantic Ocean, only seismicity from the Mid-Atlantic Ridge and a few small tectonic convergent zones have clear paths to the hydrophones. The low number of associations actually represents a large fraction of these REB events (40–60%). This percentage is higher than at Wake Island even though the distribution of events with distance is comparable (Fig. 6). This suggests that events with clear paths to Ascension have better ocean coupling properties, on average, than those visible to Wake. In addition, *T* phases recorded at Ascension tend to have energy at higher frequencies compared to those recorded at Wake. *T* phases at Ascension will often have energy up to the Nyquist frequency (60 Hz) whereas those at Wake Island rarely have energy above 40 Hz. This is a further indication that events in the Atlantic have a higher coupling efficiency. This may be due to the depth of the SOFAR axis nearly coinciding with the Mid-Atlantic Ridge (MAR) in the Northern Atlantic (NORTHROP and COLBORN, 1974) allowing the shallow earthquakes that occur on the MAR to efficiently couple energy into high frequency modes (HANSON, 1998; DE GROOT-HEDLIN *et al.*, 1998).

3.4. Point Sur, California

The hydrophone station at Pt. Sur (PSUR) consists of a single hydrophone. It is operated by the U.S. Navy and is *not* part of the IMS network. However, it has been a consistent data provider to the PIDC and is similar to the proposed IMS stations. The data from Pt. Sur has been extremely useful in developing and tuning the hydroacoustic processing. The hydrophone is located 45 km off the northern Californian coast. It is similar to the hydrophones at Wake and Ascension Islands except it is bottom mounted at 1372 meters depth, and the waveforms are sampled at 200 Hz.

Pt. Sur observes *T* phases for 11% of the events with predicted clear paths. We do not expect all of the events to produce *T* phases, however our experience from Wake suggests that at least 25% of Pacific events do. A majority of the seismicity around the Pacific Rim occurs along the western edge and thus at far distances from Pt Sur (Fig. 6). This increase in the distance to events, and the corresponding increase in attenuation, probably accounts for the lower *T* phase association rate at Pt. Sur. The depth of the hydrophone also is likely to decrease its sensitivity because the SOFAR channel axis is at ~600 meters, substantially above the hydrophone (NORTHROP, 1974).

3.5. Victoria Island, British Columbia

This station (VIB) is one of the proposed IMS *T*-phase stations and consists of a seismometer located on the island near the Pacific Coast. The sensor is a vertical component seismometer digitally sampled at 100 Hz. It has been in the current PIDC network since the start of hydroacoustic processing. The background noise at VIB tends to be very high, and there are many short transients of unknown origin. This noise contributes to the poor performance of the station, although it is probably not unusual for a seismic station located near shore.

The population of REB events predicted to have clear paths to VIB is nearly identical to that at Pt. Sur (Fig. 6), however VIB only observes 2% of these as opposed to 11% at PSUR. In other words, VIB is a factor of 5 less sensitive than PSUR, and it may actually be worse. The rate of "noise" bursts at VIB makes it likely that they will sometimes become mistakenly associated to events. This is exacerbated by their close resemblance to *T* phases.

3.6. Great Barrier Island, New Zealand

The New Zealand hydrophone station consisted of two hydrophones (NZL01 and NZL06) located 140 km northeast of Auckland off the coast of the Great Barrier Island. It was used at the PIDC for over a year, and is not part of the proposed IMS network. The hydrophones' locations (shallow water, proximity to a shipping lane) severely limits the station's performance. However, considering the monetary savings a station like this may provide, it is useful to evaluate it relative to the other stations.

The data from NZL conspicuously lack *T* phases. Only 1 *T* phase for every 3 days of data is associated to an event in the REB, and most of these are from nearby earthquakes. The *T* phases that are seen at NZL have a high frequency content starting at 10–20 Hz instead of 2 Hz typically seen at hydrophones located in the sound channel. The hydrophones at NZL are in shallow water (250 meters) which greatly attenuates energy below 20 Hz (JENSEN *et al.*, 1994). Figure 7 shows the

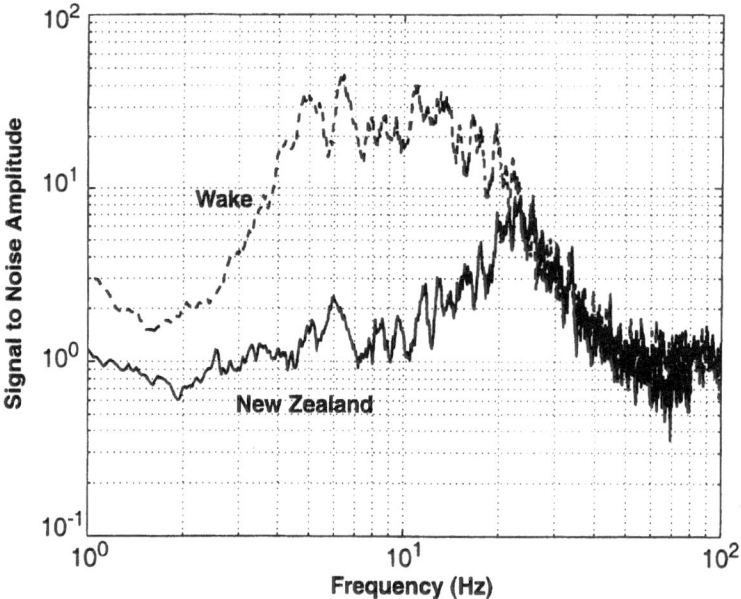

Figure 7

T-phase power spectra seen at Wake and New Zealand. The *T* phase emanates from an earthquake off the coast of Peru and contains unusually high frequency energy. The low frequencies (below 20 Hz) are attenuated at the New Zealand hydrophone relative to the Wake Island station. This is due to the shallow water (~250 meters) location of NZL.

signal-to-noise amplitude spectrum of a T phase recorded at NZL and Wake. The T phase is from an event which occurred off the coast of Peru. Both recordings contain the same high frequency energy, however NZL begins to lose energy at frequencies below 20 Hz. The low frequency energy becomes cut off at the continental shelf that extends off the coast of New Zealand. T phases generally have peak frequencies in the 2–10 Hz band, and only a few T phases have the high frequency energy necessary to be seen at NZL.

3.7. Performance of Automatic Processing

The number of detections per day varies between stations (Table 4). The average number of detections – which have been corrected for station downtime – range from 24 to 104 per day. These variations are due to factors such as ambient noise, local biological and man-made transient signals and the location of the station relative to seismically active oceanic regions.

The performance of all the stations is remarkably similar even though the T phase association rate in the REB varies significantly (Table 5). The automatic processing of data from PSUR and Wake correctly classifies and associates T phases at nearly identical rates even though Wake observes more than twice the number of T phases and 5 times the number of N phases. The T-phase station VIB misses more detections than the other stations, nonetheless the phase identification is quite comparable. Both NZL and VIB detect a lower percentage of the REB T phases associated to them than the deep water hydrophone stations. This is probably due to the high noise at these stations. However, VIB's identification rate is comparable to Ascension's. The phase identification performance at NZL is very poor because the low frequency energy in the signal, which is a major T-phase discriminant, does not propagate through the shallow water (see Section 3.6). Once the automatic system detects and correctly identifies a T phase, there is a 65% to 75% chance it will associate with the event that eventually makes it into the REB. This percentage seems not to depend on station type, which is expected since the association is only based on phase type and travel-time residual. Because the association rate does not depend on the distance to events, we can conclude that the majority of error in the arrival time is controlled by the event's location and/or coupling error and not the travel-time models.

4. Special Studies

4.1. Travel-time Table Validation

Currently at the PIDC theoretical hydroacoustic arrival times are determined from travel-time tables that are calculated using path and seasonal dependent effects (see Section 2.5). Although these travel times are calculated from well tested databases, little data exist to validate the PIDC implementation. Hydroacoustic propagation modeling is a very mature field, however predicting travel times for the IMS hydroacoustic network is made difficult by several factors. The first factor is the scope of interest. All portions of the world's oceans are of equal importance, therefore a uniform and global propagation model must be used. Another factor is that the hydroacoustic network is sparse. There is little redundancy, and station coverage for many parts of the oceans will be less than ideal. In addition, the stations are not designed for true array processing. This makes it difficult to know which ray path or mode equivalent is being measured and therefore we must predict which mode to use to estimate travel times. Ideally, we would like to use ground-truth data to test the location ability of the current system. The coverage by the current network has not yielded any in-water events that were recorded by enough hydrophone stations to estimate locations. However, an unexpected set of events in the Pacific Ocean has provided accurate ground truth to compare measured and predicted travel times between Japan and the two stations at Wake Island and Pt. Sur, California. These occurred prior to standard processing, however the PIDC was receiving continuous data from the two stations.

Between September 6 and 10, 1996, 141 dynamite charges were detonated for a seismic refraction survey off the Pacific coast of northern Japan. The source characteristics, origin times, and locations of the explosions were provided to us by Ryota Hino of Tohoku University, Japan (pers. comm.). All but three of the explosions had ground-truth locations accurate to within 100 m. We processed the data using the current PIDC system. Of the 138 accurately-located shots, 135 were detected at Wake and 119 were detected at PSUR. Epicenters could not be estimated because three stations are required to locate an event.

We estimate theoretical arrival times using the ground-truth locations and travel times from a constant velocity model and the 2-D, seasonal, travel-time tables used at the PIDC. The theoretical arrival times are compared with the automatically picked arrival times. Histograms of the travel-time residuals (observed minus predicted travel times) are shown in Figure 8. The theoretical arrival times are generally early compared to the observed arrival times regardless of which model is used. However, the seasonal tables are more accurate than the constant velocity model. The theoretical arrivals are 7 ± 1 s earlier than the observed arrivals at WAKE and 36 ± 3 s earlier at PSUR using the constant velocity model. The seasonal tables result in small time residuals, 3 ± 1 s at WAKE and 3 ± 3 s at PSUR.

The travel-time accuracy noticeably improved with the use of the travel-time tables, particularly for arrivals at PSUR. This is expected because the path to PSUR is long and passes through the northern Pacific which tends to have lower sound speed velocities (NORTHROP, 1974). The predicted travel times clearly benefit from using path-dependent travel-time tables for each station, rather than a constant velocity model. Because the explosions take place in nearly the same location, the paths between the source and receiver are nearly identical for each explosion. The spread in the travel-time residuals is likely due to the random error in measuring

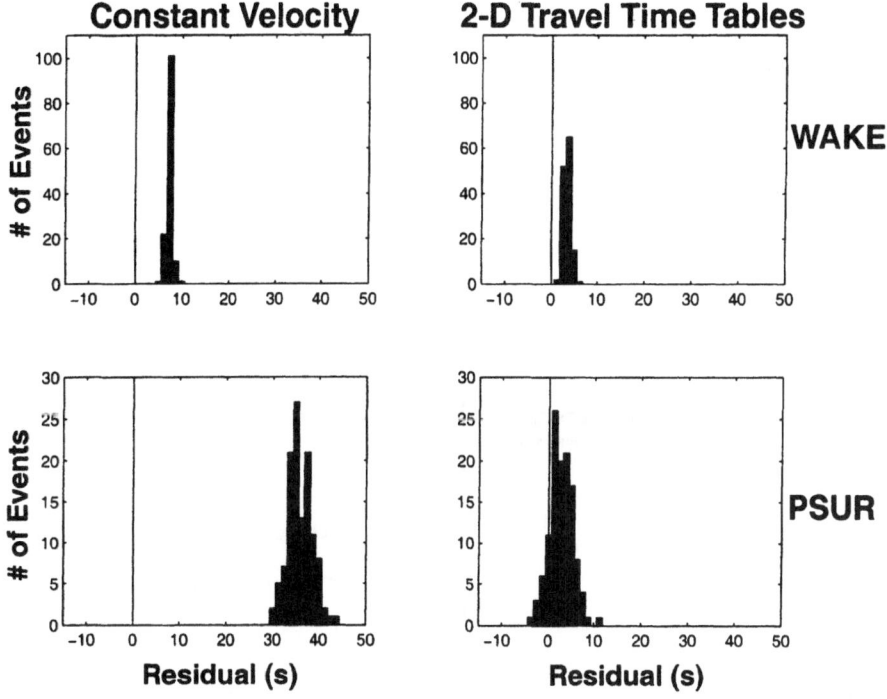

Figure 8
Travel-time residuals from the 1996 refraction experiment. The 2-D travel-time tables significantly improve
the residuals at PSUR.

the arrival time. This is further confirmed by the increase in the variance at PSUR compared to WK30. The path to PSUR is much longer, thus the signal has gone through more dispersion and has a lower SNR. The 2-sigma standard deviation is 2 and 6 seconds at WK30 and PSUR, respectively. A likely upper bound for the total error (measurement plus model) is about 12 seconds. This translates into a maximum 18 km error in location (assuming there are signals recorded on three well separated instruments).

The 1996 French Polynesian underground nuclear tests provide another set of ground truth from which to validate the travel-time tables. We have data recorded at PSUR for all six of the tests, but data for only one explosion at WK30. Ground truth was obtained from the Nuclear Explosion database available at the PIDC (BONDÁR et al., 1998). The measured and predicted travel times are shown in Figure 9. Most of the measured times fall within the predicted modeling error, however those that do not are still within 2 to 5 seconds of the predicted times. The small number of events make statistical argument difficult, but the general spread in errors is consistent with those observed from the refraction experiment.

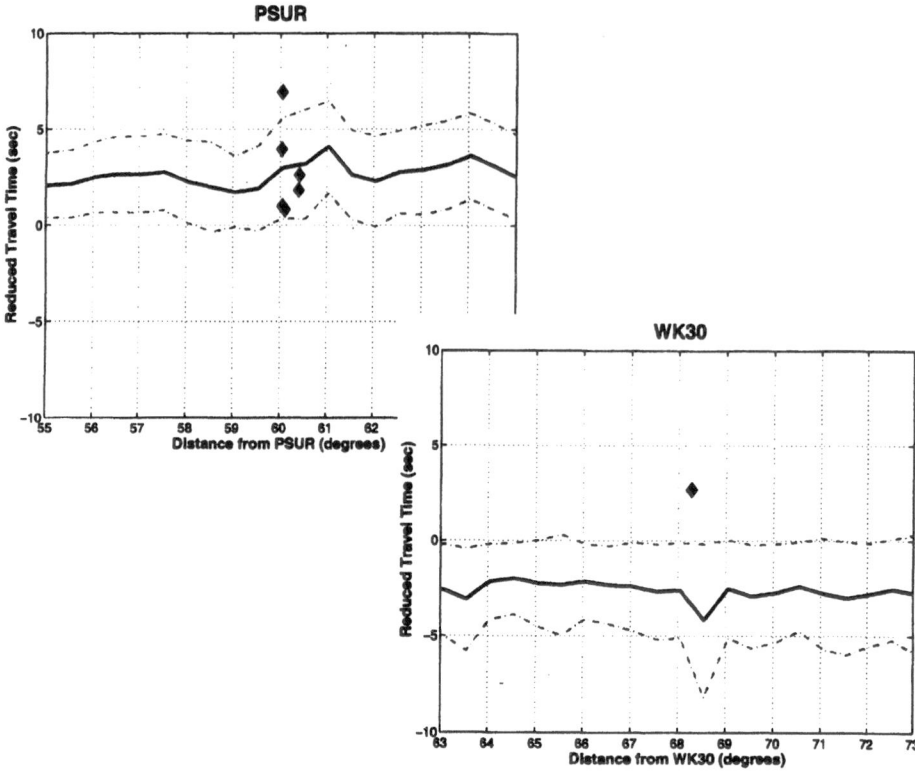

Figure 9
Comparing measured and predicted arrival times of hydroacoustic signals from French Polynesian Atoll Tests recorded at Pt. Sur and Wake Island. The solid line is the predicted time for the 2-D model, using a reduction velocity of 1.485 km/s. The dashed lines are the predicted modeling error, and the diamonds are the automatically measured arrival times.

4.2. Bubble Pulse Delay Time Estimation

An in-water explosion creates a gas bubble that expands until the ambient pressure overcomes the internal energy of the gas and forces the bubble to collapse (COLE, 1948). Near the bubble's minimum radius another shockwave is produced, and the bubble expands again. This cycle can repeat many times. The resulting pressure source function consists of a series of pulses. The delay time between the pulses is a function of the source yield and depth. Evidence of a bubble pulse in a signal is a strong indication of an in-water impulsive source and can be used to constrain some source parameters. Explosions may not, in certain cases, form a bubble or the bubble's signature can be masked by other effects, and therefore the lack of a bubble pulse does not rule out an in-water source.

There are two general methods of estimating the existence and delay time of a bubble pulse: autocorrelation and cepstral analysis (BARRODALE *et al.*, 1984). The autocorrelation produces secondary peaks at delay times corresponding to the bubble pulse. Cepstral analysis attempts to estimate any scalloping in the spectrum that would be caused by a multiple delta-like source function. It also provides a convenient method to empirically separate much of the propagation's Green's function from the source function.

The cepstrum is essentially the Fourier Transform of the signal's log power spectrum. The signal measured on the hydrophone is a convolution of the source function, the propagation Green's function and the instrument's response. In the frequency domain the convolutions become multiplications, and taking the log converts these to additions. It is assumed that the portion of the log power spectrum due to one or more bubble pulses is a constant-amplitude, periodic, zero-mean function of frequency. In that case, any broadband trend in the power spectrum is due to other source or propagation effects. The broadband trend is estimated by smoothing the log power spectrum. It is removed prior to computing the cepstrum. This reduces power in the cepstrum at short delay times which otherwise could interfere with the bubble pulse estimation. The cepstrum is processed using an iterative filtering technique to emphasize spikes within a noisy signal. It is possible that for very short delay times (i.e., broadband scalloping) the power spectrum trend will contain the bubble pulse information. To assure this is not a problem, the cepstrum of the spectral trend is also computed and stored in the database. Three values for each cepstrum are measured: delay time, cepstral variance and peak level. The delay time is estimated by fitting a quadratic around the cepstrum's maximum peak. The variance in the cepstrum is measured after removing the maximum peak. Finally, the peak level is measured relative to the variance in the cepstrum.

Two data sets are used to demonstrate the effectiveness of this algorithm. The series of explosions from the refraction experiment off the coast of Japan (see Section 4.1) provide a large set of similar sources recorded at two widely separated hydrophones (>7000 km). The other data set is from one of the Chase explosions conducted by the U.S. Navy in 1970 (PULLI et al., 1998; WEINSTEIN, 1968). This was a large explosion (an approximate yield of 600 tons of TNT equivalent) at a depth of 170 meters, and is the example most similar to an underwater nuclear explosion in our possession. These explosion data were recorded on several hydrophones at Ascension Island. In addition to the direct shock, signals that reflected off continental boundaries are observed resulting in a total of 11 detected signals.

The cepstral features for the two data sets are computed using the automatic algorithm. Figure 10 shows the cepstrum peak level versus delay time for the refraction experiment, the Chase explosion and some miscellaneous T and N phases for comparison. The delay times measured for the Chase experiment are for the most part just above 1000 milliseconds. The peak levels cover a wide span, many of which cannot be separated from the noise. The delay times from the refraction experiment range between 140 and 240 milliseconds. Some of this variation is due to different charge yields, but the delay time for charges with the same yield can vary by 30

Cepstral Measurements

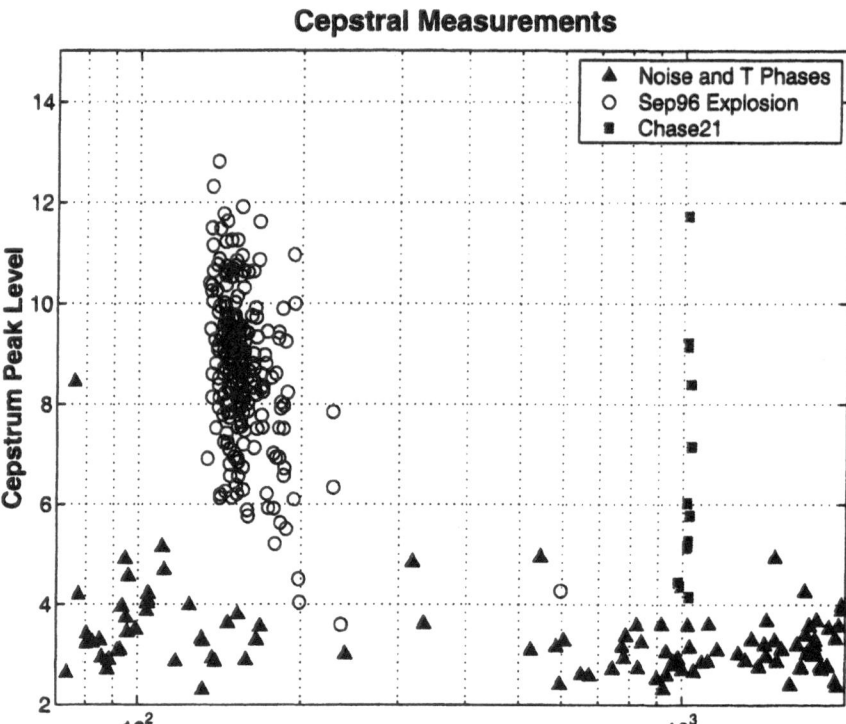

Figure 10

Cepstral measurements for signals from the 1996 refraction experiment off the coast of Japan and the large Navy explosion in 1970 (Chase 21). The cepstral peak level is measured relative to the background cepstrum.

milliseconds. The peak levels are better separated from the noise than was the case for the Chase data. However the two largest explosions (400 kg) have low peak values, indicating the folly of relying too heavily on cepstral values for identifying the source.

The refraction explosions were recorded at two stations (WK30 and PSUR) which are at vastly different distances and have completely separate recording instrumentation. This provides two independent measurements of the cepstrum which can be used to estimate the precision of the delay time measurements. Using the automatic results, we differenced the delay time between the two stations for each explosion detected. There were two large outliers due to sub-optimal onset and termination time estimations caused by interfering signals. Figure 11 is a histogram of the remaining 95 difference pairs. Although the variation in delay times among the different explosions is around 100 milliseconds, the differences between PSUR and WAKE for the same explosion are less than 2 milliseconds. The average difference is

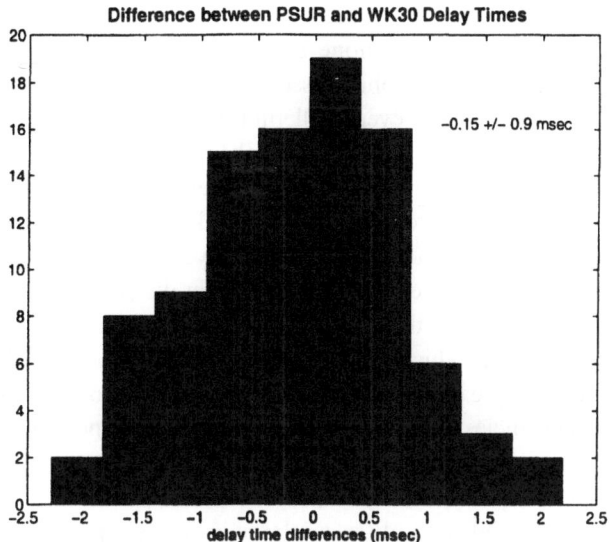

Figure 11

Differences in measured bubble pulse delay times between signals recorded at PSUR and WK30. The paths to the two stations are very different as is the recording instrumentation, and the errors in the two measurements can be considered statistically independent. The sub-sample rate accuracy was achieved by interpolating around the peak in the cepstrum.

essentially zero with a standard deviation of less than 1 millisecond. The sample rates at PSUR and WK30 are 200 Hz and 320 Hz, respectively (or a sample spacing of 5 and 3 milliseconds). Obtaining delay times with sub-sample rate precision is possible because we interpolate about the cepstral peak.

5. Conclusion

A hydroacoustic processing system has been designed, built and implemented at the PIDC. It follows similar steps to those used in seismic processing, but the unique source generation, propagation medium, and network properties require specialized units within each of the general steps. Waveform features are measured for each detection over a suite of frequency bands and are used to classify signals into one of three categories (H, T or N). This is accomplished using either a neural network or a set of standard rules. The neural network improves the results, but requires the existence of a training database. The T and H detections are combined with the seismic and infrasonic data to form an automatic event bulletin that is then reviewed by human analysts before the final bulletin is produced. T phases are only associated to events formed by the seismic network, however H phases can also be used in locating events. Currently associations are based on phase identification and travel time only. An algorithm is in the process of being implemented that will produce bearing estimates from the hydrophone stations with multiple sensors. This will further increase confidence in the phase associations.

The REB is the largest global event bulletin that associates hydroacoustic phases on a routine basis. The hydroacoustic phases in the reviewed event bulletin (REB) are primarily T phases. This is for two reasons. Deep in-water explosions are rare occurrences, and the current hydroacoustic network has too few stations to locate events in most ocean areas. 25% of REB events have associated T phases, and they show great potential for use as earthquake/explosion discriminants. The number of T phases seen at a hydrophone depends on the station's proximity to seismicity and the region's efficiency at producing T phases. Shallow water hydrophones are not very useful for recording earthquake generated T phases, however they do record energy at higher frequencies which in-water explosions are expected to produce. The only T-phase station in the current PIDC network records earthquake T phases rather poorly compared to typical hydrophone stations, but we do not have data to determine the station's ability to observe a large in-water explosion.

Specific studies have been conducted to evaluate various aspects of the hydroacoustic system. The 2-D seasonal travel-time tables produce highly accurate results over paths where ground-truth data is available. With adequate coverage, locations accurate to within several kilometers should be obtainable. The cepstral analysis used at the PIDC produces highly precise results. The existence of a strong cepstral peak is a strong indicator of an in-water source, but the absence of a peak cannot rule out an in-water source. The accuracy of the delay time estimation (< 2 milliseconds) could be useful in linking signals among stations.

Appendix

Probability Weighted Time

The probability weighted time is an estimation of the peak energy arrival time. The method can be applied to any signal type, although currently is only applied to hydroacoustic data to estimate arrival times of the peak energy.

A.1. Peak Probability

Consider a set of samples from a time series $(x_1, \ldots, x_N)$ which have values $(\lambda_1, \ldots, \lambda_N)$ and a sampling interval Δt. We wish to find the time corresponding to the maximum value of the series. However, each value has an uncertainty associated with it, and thus the maximum of λ_i may not be the maximum of x_i. It is assumed that each sample has the same uncertainty, σ. Then if the errors are Gaussian distributed, the probability that the true value of x_i lies in some range of width dx_i is

$$P(x_i)dx_i = \frac{1}{\sqrt{2\pi\sigma^2}} e^{(x_i - \lambda_i)^2/(2\sigma^2)} dx_i \ .$$

It follows that the probability that x_i is less than some value, y, given its measured value, λ_i, is

$$P(x_i < y) = \frac{1}{\sqrt{2\pi\sigma^2}} \int_{-\infty}^{y} e^{(x_i - \lambda_i)^2/(2\sigma^2)} dx_i = \mathrm{erf}\left(\frac{y - \lambda_i}{\sigma}\right) \ .$$

From this we can calculate the probability that a particular x_n is the true peak.

$$P_n \equiv P(x_n \text{ is the peak}) = \int_{\substack{(\text{all possible} \\ \text{values of } x_n)}} \left(\begin{array}{l} \text{probability of any} \\ \text{particular value } x_n \end{array}\right)$$

$$\times \prod_{\substack{(\text{all other } x_i) \\ i \neq n}} \left(\begin{array}{l} \text{probability} \\ \text{that } x_i < x_n \end{array}\right)$$

$$P_n = \frac{1}{\sqrt{2\pi\sigma^2}} \int_{-\infty}^{\infty} dx_n e^{(x_n - \lambda_n)^2/(2\sigma^2)} \prod_{i \neq n} \mathrm{erf}\left(\frac{x_n - \lambda_i}{\sigma}\right) \ .$$

The behavior of the above equation becomes more intuitive if we substitute variables: $x = (x_n - \lambda_n)/\sigma$

$$P_n = \frac{1}{\sqrt{2\pi}} \int_{-\infty}^{\infty} dx\, e^{-x^2/2} \prod_{i \neq n} \mathrm{erf}\left(\frac{x + \lambda_n - \lambda_i}{\sigma}\right) \ .$$

Note that the second term in the argument of the erf is simply the number of standard deviations by which two measured values differ. If $\lambda_j < \lambda_k$, the erf in P_j will have smaller magnitude arguments than the corresponding arguments for P_k, and hence $P_j < P_k$. Also, if there is some λ_k; $(\lambda_k \gg \lambda_j)$ for all $(j \neq k)$, the arguments in the erf for P_k are positive, and the values of the erf approach 1. For all other P_j, the arguments in the erf are negative, and their values approach 0. Hence $P_k \to 1$, while all other $P_j \to 0$.

A.2. Arrival Time Estimation

To obtain the probability function for the time series an estimate of the uncertainty is needed. Currently, in hydroacoustic processing, the average predetection noise level is used as the uncertainty in measured values. To estimate the most likely time of arrival of peak energy, we compute a time average of $P(t_i)$,

$$\text{Probability weighted time} = \langle t \rangle_P = \sum_n P_n t_n$$

$$= t_0 + \Delta t \sum_n P_n n \ .$$

Note that if two samples are near to one another in intensity, the corresponding P_i will be nearly equal, and $\langle t \rangle_P$ will be an average between the samples. Likewise, if sample j is clearly dominant, $P_j \approx 1$ and $\langle t \rangle_P \approx t_j$.

The variance in the time estimate is,

$$\sigma^2 \langle t \rangle_P = \sum_n (P_n t_n^2 - \langle t \rangle_P^2) \ .$$

In the case with a high SNR (i.e., small uncertainty), if any one sample i dominates, then $P_i \to 1$, and the measurement error is very small. By contrast, for low SNR, the values, P_n, are more evenly distributed, and the measurement error grows large. Identical waveforms may have different measurement errors if they have different noise levels.

REFERENCES

ANGELL, J., FARRELL, T., and PULLI, J. (1998), *Characterization of Reflected Hydroacoustic Signals*, Proceedings of the 20th Annual Seismic Research Symposium on Monitoring a Comprehensive Test-Ban Treaty, 666–675.

BACHE, T. C., BARKER, T. G., BROWN, M. G., PYATT, K. D., and SWANGER, H. J. (1980), *The Underwater Signature of a Nuclear Explosion at the Ocean Surface*, Technical Report VSC-TR-81-24, VELA Seismological Center.

BARRODALE, I., CHAPMAN, N. R., and ZALA, C. A. (1984), *Estimation of Bubble Pulse Wavelets for Deconvolution of Marine Seismograms*, Geophys. J. R. astr. Soc. 77, 331–341.

BISHOP, C. M., *Neural Networks for Pattern Recognition* (Oxford University Press, New York, 1995).

BONDÁR, I., YANG, X., WANG, J., BAHAVAR, M., ISRAELSSON, H., and McLAUGHLIN, K. (1998), *Tuning and Calibration Activities at the PIDC*, Proceedings of the 20th Annual Seismic Research Symposium on Monitoring a Comprehensive Test-Ban Treaty, 1–10.

BRATT, S. R. (1996), *The CTBT International Data Centre*, EOS 77, F3.

BROWN, D. J., KATZ, C. N., LE BRAS, R., WANG, J., and GAULT, A. (1998), *Infrasonic Processing at the Prototype International Data Center*, Proceedings of the 20th Annual Seismic Research Symposium on Monitoring a Comprehensive Test-Ban Treaty, 555–562.

BURGER, M. D., BOUCHER, C. E., DALEY, E. M., RENNER, W. W., PASTOR, V. L., HAINES, L. C., and ELLER, A. I. (1994), *Acoustic System Performance Model ASPM 4.0A: Users Guide*, SAIC-94/1000, Science Applications International Corporation, McLean, VA.

CHAPMAN, N. R. (1985), *Measurement of the Waveform Parameters of Shallow Explosive Charges*, J. Acoust. Soc. Am. 78, 672–681.

CLARKE, D. B., HARBEN, P. E., ROCK, D. W., and WHITE, J. W. (1997), *Energy Coupling of Nuclear Burst in and above the Ocean Surface: Source Region Calculations and Experimental Validation*, Proceedings of the 19th Annual Seismic Research Symposium on Monitoring a Comprehensive Test-Ban Treaty, 723–731.

COLE, R. H., *Underwater Explosions* (Princeton University Press. Princeton, New Jersey, 1948).

DE GROOT-HEDLIN, C., BLACKMAN, D., and ORCUTT, J. (1998), *Observations and Numerical Modeling of T-phase Coda*, Proceedings of the 20th Annual Seismic Research Symposium on Monitoring a Comprehensive Test-Ban Treaty, 657–665.

DUDA, R. O., and HART, P. E., *Pattern Recognition and Classification* (John Wiley, New York, 1973).

FOX, C. G., RADFORD, E., DZIAK, R. P., LAU, T., MATSUMOTO, H., and SCHREINER, A.E. (1995), *Acoustic Detection of a Seafloor Spreading Episode on the Juan de Fuca Ridge Using Military Hydrophone Arrays*, GRL *22*, 131–134.

GITTERMAN, Y., BEN-AVRAHAM, Z., and GINZBURG, A. (1998), *Spectral Analysis of Underwater Explosions in the Dead Sea*, Geophys. J. Inter. *134*, 460–472.

HANSON, J. A. (1998), *Seismic and Hydroacoustic Investigations near Ascension Island*, Ph.D. Thesis, University of California, San Diego.

HANSON, J. A., GIVEN, H. K., and HARRIS D. (1997), *Performance of an Island Seismic Station for Observing Hydroacoustic T Phases*, EOS *78*, F478–479.

HANSON, J. A., and GIVEN, H. K. (1998), *Accurate Azimuth Estimates from a Large Aperture Hydrophone Array Using T-phase Waveforms*, GRL *25*, 365–368.

HORNIK, K. (1991), *Approximation Capabilities of Multilayer Feedforward Networks*, Neural Networks *4*, 251–257.

JENSEN, F. B., KUPERMAN, W. A., PORTER, M. B., and SCHMIDT, H., *Computational Ocean Acoustics* (AIP Press, Woodbury, New York, 1994).

KURKOVA, V. (1991), *Kolmogorov's Theorem and Multilayer Neural Networks*, Neural Networks *5*, 501–506.

LANEY, H. (1994), *The Fast Broadband Interpolation Method: A New Algorithm for Approximating Broadband Acoustic Phenomena Using Normal Modes*, M.A. Thesis, University of Maryland, College Park.

LANEY, H., DYSART, P., FREESE, H., and WILLEMANN, R. (1996), *An Automated System for Detecting and Classifying In-water Explosions and T Phases*, J. Acoust. Soc. Am. *100*, 2641.

LAVERGNE, M. (1970), *Emission by Underwater Explosions*, Geophysics *35*, 419–435.

LAY, T., and WALLACE, T. C., *Modern Global Seismology* (Academic Press, San Diego, 1995), pp. 521–523.

MCCREERY, C. S., and WALKER, D. A. (1988), *The Wake Island Hydrophone Array*, Seis. Res. Lett. *59*, 22.

MILNE, A. R. (1959), *Comparison of Spectra of an Earthquake T Phase with Similar Signals from Nuclear Explosions*, BSSA *49*, 317–329.

NORTHROP, J. (1974), *SOFAR Accuracy in the North Pacific*, J. Acoust. Soc. Am. *55*, 191–193.

NORTHROP, J., and COLBORN, J. G. (1974), *SOFAR Channel Axial Sound Speed and Depth in the Atlantic Ocean*, JGR *79*, 5633–5641.

OKAL, E., and TALANDIER, J. (1997), *T Waves from the Great 1994 Bolivian Deep Earthquake in Relation to Channeling of S-wave Energy up the Slab*, JGR *102*, 27,421–27,437.

PISERCHIA, P-F.,VIRIEUX, J., RODRIGUES, D., GAFFET, S., and TALANDIER, J. (1998), *Hybrid Numerical Modelling of T-wave Propagation: Application to the Midplate Experiment*, Geophys. J. Inter. *133*, 789–800.

PORTER, M. B., and REISS, E. L. (1985), *A Numerical Method for Bottom Interacting Ocean Acoustic Normal Modes*, J. Acoust. Soc. Am. *77*, 1760–1767.

PULLI, J., DORFMAN, E., FARRELL, T., and ANGELL, J (1998), *Modeling Hydroacoustic Waveform Envelopes Using Normal Modes*, Proceedings of the 20th Annual Seismic Research Symposium on Monitoring a Comprehensive Test-Ban Treaty, 689–697.

RINGDAL, F. (1996), *Developing and Testing an Experimental International Seismic Monitoring System; the GSETT-3 experiment*, EOS *77*, F4.

RUMELHART, D. E., and MCCLELLAND, J. L., *Learning Internal Representations by Error Propagation in Parallel Distributed Processing* (Vol. 1, MIT Press, Boston, 1986).

SHOCKLEY, R. C., NORTHROP, J., and HANSEN, P. G. (1982), *SOFAR Propagation Paths from Australia to Bermuda: Comparison of Signal Speed Algorithms and Experiments*, J. Acoust. Soc. Am. *71*, 51.

STEVENS, J. L., BAKER G. E., MURPHY, J. R., COOK, R. W., D'SPAIN, G., BERGER, L. P., and KHRISTOFOROV, B. D. (1998), *T-phase Excitation and Transfer Function Research*, Proceedings of the 20th Annual Seismic Research Symposium on Monitoring a Comprehensive Test-Ban Treaty, 698–707.

TALANDIER, J., and OKAL, E. (1998), *On the Mechanism of Conversion of Seismic Waves to and from T Waves in the Vicinity of Island Shores*, BSSA *88*, 621–632.

URICK, R. J., *Principles of Underwater Sound*, 3rd Edition (McGraw-Hill, New York, 1983).

WEINSTEIN, M. S. (1968), *Spectra of Acoustic and Seismic Signals Generated by Underwater Explosions during the Chase Experiment*, J. Geophys. Res. *73*, 5473–5476.

WILLEMANN, R. (1998), *Hydroacoustic Detection, Phase Identification, Event Location at the IDC*, Proceedings of the 19th Annual Seismic Research Symposium on Monitoring a Comprehensive Test-Ban Treaty, 12–13.

(Received June 29, 1999, revised December 17, 1999, accepted January 3, 2000)

 To access this journal online:
http://www.birkhauser.ch

Pure appl. geophys. 158 (2001) 457–474
0033–4553/01/030457–18 $ 1.50 + 0.20/0

⎰Pure and Applied Geophysics

Converted T Phases Recorded on Hawaii from Polynesian Nuclear Tests: A Preliminary Report

EMILE A. OKAL[1]

Abstract — We present a preliminary study of T waves from Polynesian nuclear tests at Mururoa, recorded on digital stations of the Hawaii Volcano Observatory network, following their conversion to seismic waves at the southern shore of the Island of Hawaii, and subsequent propagation to the recording stations. We show that seismograms are composed of several packets, which can be interpreted as resulting from $T \rightarrow P$ and $T \rightarrow S$ conversions, and which feature distinct spectral characteristics. As the distance from the shoreline to the station increases, the relative importance of the several wave packets changes; a prominent shadow for $T \rightarrow P$ is found at 8–12 km from the shore. This pattern is affected by the local crustal structure; in a favorable case, propagation in deep, low-attenuation layers resulted in a clear record as far as 76 km from the shoreline. While these results are generally robust, they can be moderately affected by a change of location of the source inside Mururoa Atoll.

Key words: T phases, seismic conversions, Nuclear Tests.

1. Introduction and Background

The purpose of this paper is to examine the seismic waves generated on the island of Hawaii (hereafter "the Big Island") upon reception of T waves generated by a number of nuclear tests at Mururoa, French Polynesia, over the years 1986–1991. Our goal is to understand the exact nature of the conversion of their energy into seismic waves, and of their subsequent propagation through the structure of the island. In the framework of the monitoring of the Comprehensive Test-Ban Treaty (CTBT), an improved understanding of these mechanisms of conversion and propagation will be necessary to adequately interpret any signals received at the so-called "T-wave stations" mandated by the treaty. In addition, this research may help optimize the deployment or relocation of T-wave stations in order to maximize their usefulness.

T waves are acoustic vibrations efficiently propagated in the SOFAR channel of the world's oceans. They were mentioned as early as 1930 (ANONYMOUS, 1930), recognized as teleseismic in nature independently by RAVET (1940) and LINEHAN

---

[1] Department of Geological Sciences, Northwestern University, Evanston, IL 60201, USA. E-mail: emile@earth.nwu.edu

(1940), and correctly described as guided waves following intense research during World War II (e.g., PEKERIS, 1948). The combination of an efficient waveguide and the practical absence of attenuation in the seawater at low acoustic (high seismic) frequencies (3–20 Hz) makes them a choice agent for the propagation (and eventual detection) of extremely small signals to extremely large distances in the marine environment.

Upon reaching the shore of an island (or continent), the acoustic energy in a T wave is converted to seismic energy, the resulting waves being then able to propagate in the island or continent structure (e.g., CANSI and BETHOUX, 1985; COOK and STEVENS, 1998), be recorded by standard seismographic stations, or even, when sufficiently powerful, felt by the local population (e.g., TALANDIER and OKAL, 1979). While this conversion is simply an illustration of the classic problem of reflection and refraction of elastic waves at a surface of discontinuity, it is made complex by a number of specific circumstances. Most importantly, the conversion is controlled by the exact geometry of the converting slope. In addition to being often unknown on the scale of the relevant wavelengths (e.g., 300 m at 5 Hz in the water), the latter is expected to depart from the simple model of a planar interface. In particular, bays and bights have been shown to provide focusing effects resulting in the preferential radiation of T-wave energy during the converse process of seismic-to-acoustic conversion near an earthquake source (OKAL and TALANDIER, 1997).

For this reason, acoustic-to-seismic conversion at island shores has been the subject of intense research, both theoretical (e.g., PISERCHIA et al., 1998) and observational. Among the latter studies, KOYANAGI et al. (1995) have investigated the propagation across the Big Island of converted T phases to infer their attenuation characteristics. However, the earthquake sources which they used (including the 1989 Loma Prieta event) involved the additional complexity of seismic-to-acoustic source-side conversion. In addition, MCLAUGHLIN (1997) has compared seismic phases converted from Mururoa T waves and recorded at San Nicolas Island, off the Southern California coast, to hydrophone records of the same events obtained at Big Sur. His study suffers however, from the lateral distance between the two receiving sites, and from the extreme complexity of the continental shelf off Southern California. Finally, PASYANOS and ROMANOWICZ (1997) have documented the propagation of converted T phases from Mururoa events as guided surface waves, up to 200 km inside Northern California.

More recently, TALANDIER and OKAL (1998; hereafter Paper I) used a variety of recording environments in Polynesia (atolls, high islands) and of T-wave sources (Hawaiian earthquakes, chemical marine explosions) to study in detail the conversion of T-wave energy to and from seismic waves in a number of case scenarios. They concluded that steep segments of ocean floor in the immediate vicinity of a shoreline are efficient converters of T waves into seismic P waves. As the slope angle decreases, the T wave becomes post-critical for a simple refraction, resulting in the development of a shadow zone, and in preferred conversion to S, and possibly guided (Rayleigh-type) waves. These results, upheld by the finite-difference calculations of PISERCHIA *et al.* (1998), strongly suggest that the optimal location of a *T*-wave station may be in the immediate vicinity of a shoreline featuring a steep slope at the depths (~1200 m) characteristic of the axis of the SOFAR channel.

This geometry can be achieved by locating the station on an atoll, where coral walls are known to offer slopes as steep as 45° to 60° (GUILLE *et al.*, 1993), but it can also be found at the head of fresh basaltic flows, comparable to aerial "palis," and well documented in the offshore bathymetry of the southern flank of the Big Island (SMITH, 1994). In this general context, the present paper gives a preliminary study of the reception of converted phases by seismic stations of the Hawaii Volcano Observatory (HVO) network, following a number of nuclear explosions at Mururoa.

2. Dataset

Figure 1 shows the general layout of the present study. Its geometry features several significant advantages: first and foremost, the use of nuclear explosions at Mururoa provides a perfect case study in the framework of the monitoring of the CTBT. The location of the test site on a Polynesian atoll with steep reefs minimizes the complexity of the conversion process at the source (we do not consider here any

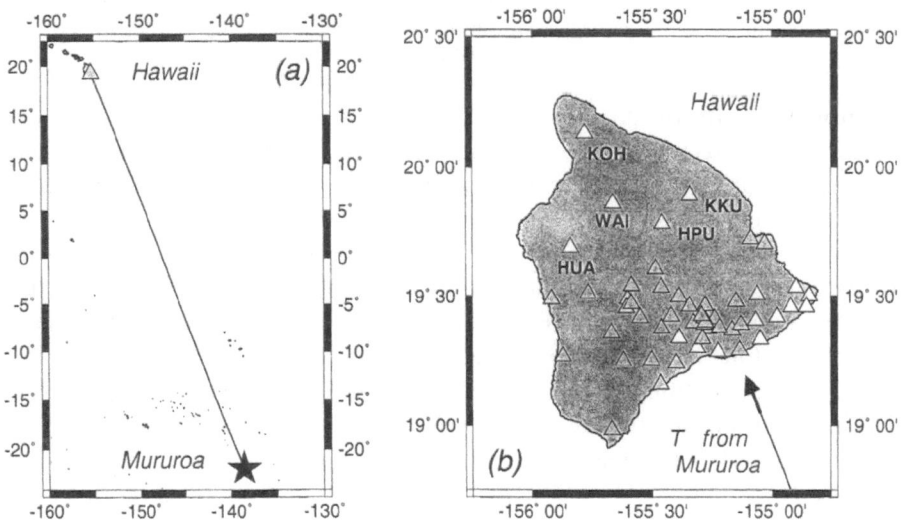

Figure 1

(a): General layout of the study. The star shows the Polynesian test site at Mururoa, and the triangle schematizes the receiving stations on the Big Island of Hawaii. *(b)*: Map of the HVO network on the Big Island. The open symbols are the stations referenced in the text (see also Table 2). Codes for southern shore stations are detailed on frame *(c)*. The large arrow identifies the azimuth of arrival of the *T* phase.

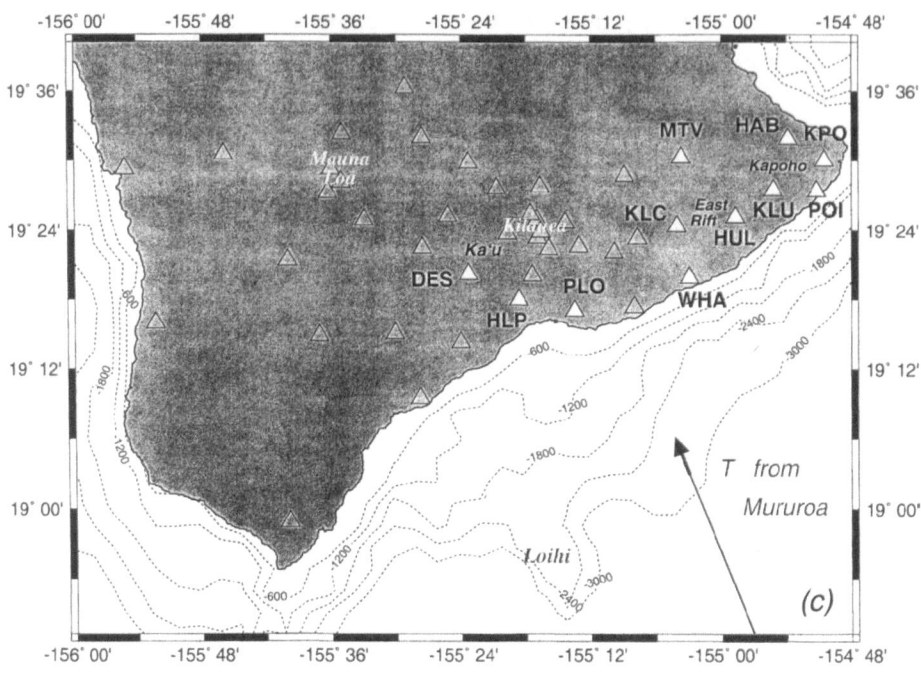

Figure 1c
Close-up of the southern shore of the Big Island, showing the stations used in this study (open triangles), and the bathymetry down to 3000 m. The 1200-m contour represents the axis of the SOFAR and is expected to control the conversion. Note the different underwater morphology off stations WHA and HLP.

tests from the southern site at Fangataufa, which entice a more complex generation process at the source). In addition, the shoreline of the Big Island is relatively regular, grossly perpendicular to the direction of arrival of the T wave, and most importantly was recently surveyed to a horizontal precision of 20 m (SMITH, 1994). Finally, HVO operates a unique, dense network of short-period seismic stations across the island, and especially on the southern flank of Kilauea Volcano, digitally recorded since 1986. Several of the stations are equipped with three-component seismometers.

While continuous digital recording is in principle available, T-wave signals have occasionally been strong enough to trigger the recording algorithm set for local Kilauea events, and are thus accessible at minimal effort from optical disks, with up to 80 channels of 100 sample per second data available.

Table 1 lists the 12 tests at Mururoa (out of a total of 28 for the period 1986–1996) which triggered local recording at HVO. Their epicentral coordinates and origin time are those listed by the National Earthquake Information Center (NEIC). In the present study, we assume that the origin time was at the exact minute, which suggests a systematic error of -1.7 ± 0.4 s in the NEIC source time. Table 2 gives a list of stations used in this study.

Table 1

Mururoa events which triggered HVO digital recording

Date D M (J) Y	NEIC Epicenter			Probable Origin Time GMT	Magnitudes		Number of stations	Peak-to-peak amp. at HUL (digital units)
	Latitude (°N)	Longitude (°E)	Origin Time GMT		m_b	M_S		
12 NOV (316) 1986	−21.911	−139.085	17:01:58.3	17:02:00	5.3		88	1776
5 MAY (125) 1987	−21.900	−139.096	16:57:57.7	16:58:00	4.9		14	835
20 MAY (140) 1987	−21.893	−138.964	17:04:58.3	17:05:00	5.6		181	2799
25 MAY (146) 1988	−21.903	−139.009	17:00:58.4	17:01:00	5.6	4.9	120	3880
23 NOV (328) 1988	−21.991	−138.907	17:00:58.0	17:01:00	5.4		99	2956
11 MAY (131) 1989	−21.853	−139.006	16:44:58.3	16:45:00	5.6		53	1021
3 JUN (154) 1989	−21.835	−138.996	17:29:58.5	17:30:00	5.3		70	1077
20 NOV (324) 1989	−21.851	−138.964	17:28:58.4	17:29:00	5.3		87	1583
2 JUN (153) 1990	−21.877	−138.918	17:29:58.5	17:30:00	5.3		96	739
4 JUL (185) 1990	−21.850	−139.042	17:59:58.7	18:00:00	5.1		69	691
18 MAY (138) 1991	−21.832	−139.014	17:14:58.5	17:15:00	5.1		71	955
15 JUL (196) 1991	−21.877	−138.963	18:09:58.3	18:10:00	5.3		119	3466

Table 2

HVO stations used in this study

Code	Station name	District	Latitude (°N)	Longitude (°E)	Distance to coast (km)
		T waves observed			
WHA	Waha'ula	East Rift	19.332	−155.049	< 1
HUL	Heiheiahulu	East Rift	19.419	−154.979	7
KLC	Kalalua Cone	East Rift	19.407	−155.068	9
MTV	Mountain View	East Rift	19.504	−155.063	18
POI	Pohoiki	Kapoho	19.457	−154.854	< 1
KLU	Pu'u Kali'u	Kapoho	19.458	−154.921	4
KPO	Kapoho	Kapoho	19.500	−154.842	4
HAB	Hawai'i Beaches	Kapoho	19.532	−154.898	12
POL	Poliokeawe	Halape–Ka'u	19.284	−155.225	2
HLP	Halape	Halape–Ka'u	19.299	−155.311	4
DES	Desert	Halape–Ka'u	19.337	−155.388	11
KKU	Keanakolu	Mauna Kea	19.890	−155.343	76
		No T wave recorded			
HPU	Hale Pohaku	Mauna Kea	19.781	−155.311	64
HUA	Hualala'i		19.688	−155.458	71
WAI	Waiki'i		19.860	−155.660	75
KOH	Kohala		20.128	−155.780	114

With the exception of Section 5, we now focus on records from the event of 20 May 1987 (origin time 17:05:00 GMT), which has both the largest body-wave magnitude ($m_b = 5.6$), and the largest number of readings reported to the NEIC. The available dataset consists of vertical seismograms at 48 stations, with horizontal channels at a subset of 12 sites; The resulting 72 triggered channels were extracted and converted to SAC format. Among the many stations available on the southern flank of the Big Island, we first study the record at Heiheiahulu (HUL). While a number of stations are located closer to the shoreline, HUL, being 7 km inland, benefits from reduced background noise (incidentally, it provided the most spectacular records of the 1994 Bolivian T waves (OKAL and TALANDIER, 1997)). This enhanced distance also allows separation of the various converted seismic phases. In addition, HUL is one of the stations most regularly included in triggered datasets, allowing direct comparison of the signals from several Mururoa sources. Finally, all three components of ground motion are available at HUL.

3. The T Phase at HUL: Identification of Wave Trains

The crustal structure of the Big Island has been the subject of numerous investigations (e.g., HILL and ZUCCA, 1987). In this study, we use HILL's (1969) model, featuring a 3-layer 11-km thick crust, and directly applicable to the southern flank of Kilauea where the conversion and short propagation to HUL take place. In the framework of Paper I, and using SMITH's (1994) bathymetric data, we then predict that the $T \rightarrow P$ conversion at HUL should be composed of three arrivals, at respectively 17:59:51.8, 17:59:52.8 and 17:59:54.0 GMT.

Figure 2 shows the arrival of the T wavetrain from Mururoa, as recorded on the digital triggered system; this is a 17-second close-up of the seismogram, after band-pass filtering between 1 and 10 Hz. Note that HVO operates on Hawaiian Standard Time (HST), which is 10 hours behind GMT; all times will be given after network trigger time (07:59:42.3 HST or 17:59:42.3 GMT). It is clear that the structure of the wave at HUL is in excellent agreement with that predicted: the T phase is clearly composed of a number of arrivals featuring different characteristics in terms of ground motion polarization and frequency content. The vertical component of the ground motion is dominated by a first high-frequency group of three arrivals, all matching the predicted times within 0.5 s, followed at 14.6 s by a larger, but lower-frequency, arrival. We interpret the former as $T \rightarrow P$ conversions (P_1, P_2, P_3), and the latter as $T \rightarrow S$. This interpretation is confirmed by spectrogram analysis, as shown on Figure 3, indicating energy maxima at 7.5 Hz for P_1, 7.0 Hz for P_2 and 6.5 Hz for P_3, but only 2.8 Hz for $T \rightarrow S$. The vertical seismogram at HUL also bears striking resemblance to the T phase shown on Figure 2 of Paper I, obtained in a reverse geometry (a Hawaiian earthquake recorded at station PMO in Polynesia). Paralleling the study in Paper I, we can characterize the difference in spectrum through the ratio

$$R_{SP} = \frac{X_{T \rightarrow P}(f_{T \rightarrow S})/X_{T \rightarrow S}(f_{T \rightarrow S})}{X_{T \rightarrow P}(f_{T \rightarrow P})/X_{T \rightarrow S}(f_{T \rightarrow P})} \tag{1}$$

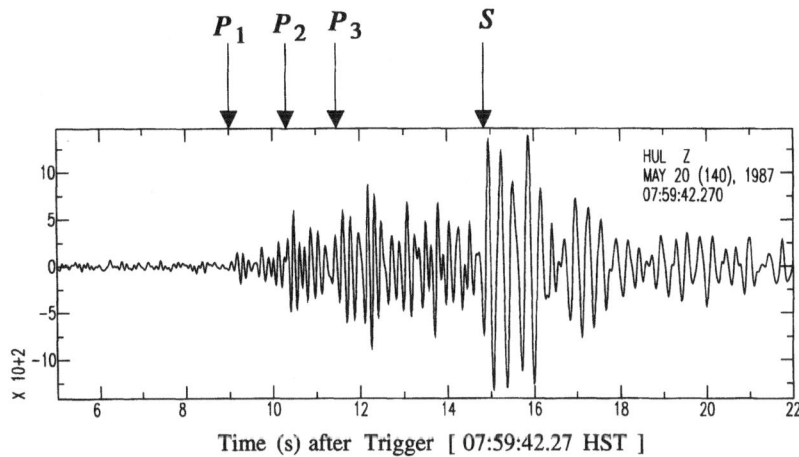

P_1 P_2 P_3 S

Figure 2

T-phase arrival at Heiheiahulu (HUL); this is a band-pass-filtered window ($1 \leq f \leq 10$ Hz) of the vertical seismogram. Note the several wave packets composing the $T \rightarrow P$ and $T \rightarrow S$ conversions.

of the relevant spectral amplitudes at the peak frequencies $f_{T \rightarrow P} = 7$ Hz and $f_{T \rightarrow S} = 2.8$ Hz of the two wave trains. When applied to two 2-s windows centered on the P_2 and $T \rightarrow S$ phases, $R = 0.025$, which if interpreted as the effect of anelastic attenuation along the path from the conversion point to HUL, yields $Q_\mu = 19$ (we assume here no bulk attenuation Q_K^{-1}). This value is in excellent agreement with that found in Paper 1 ($Q_\mu = 21$) for the reverse situation along a path similarly sampling the flank of Kilauea 15 km to the West, and also generally consistent with the value ($Q_\mu = 30$) proposed by KOYANAGI et al. (1995) in the immediate vicinity of the crater. Such values of Q would be regarded as very low in classical seismology, but are indeed documented in the particular setting of an active volcanic structure (AKI et al., 1977; TALANDIER and OKAL, 1996).

Figure 4a shows the horizontal ground motion of the presumed $T \rightarrow S$ conversion, for the window comprised between 14.5 and 16.5 s after trigger time. The polarization of the wave train is in the azimuth N120°E. While this could correspond to SV polarization for a seismic S wave converted slightly off the great circle path (the backazimuth from HUL to Mururoa is N157°E), it could also stem from diffraction at a preferential conversion site such as an underwater bay or cove; further interpretation of the wave's polarization would be speculative at this point.

The later part of the horizontal ground motion is dominated (between 18 and 21 s following trigger time) by a wave of relatively low frequency (3 Hz). As shown on Figures 4b and 4c, this wave is polarized practically north–south and its vertical and NS components feature a $\pi/2$ phase offset. This is characteristic of a surface- or interface-wave; the timing of this arrival, suggestive of a group velocity from the conversion point to HUL of only 0.55 km/s, also supports this interpretation; the

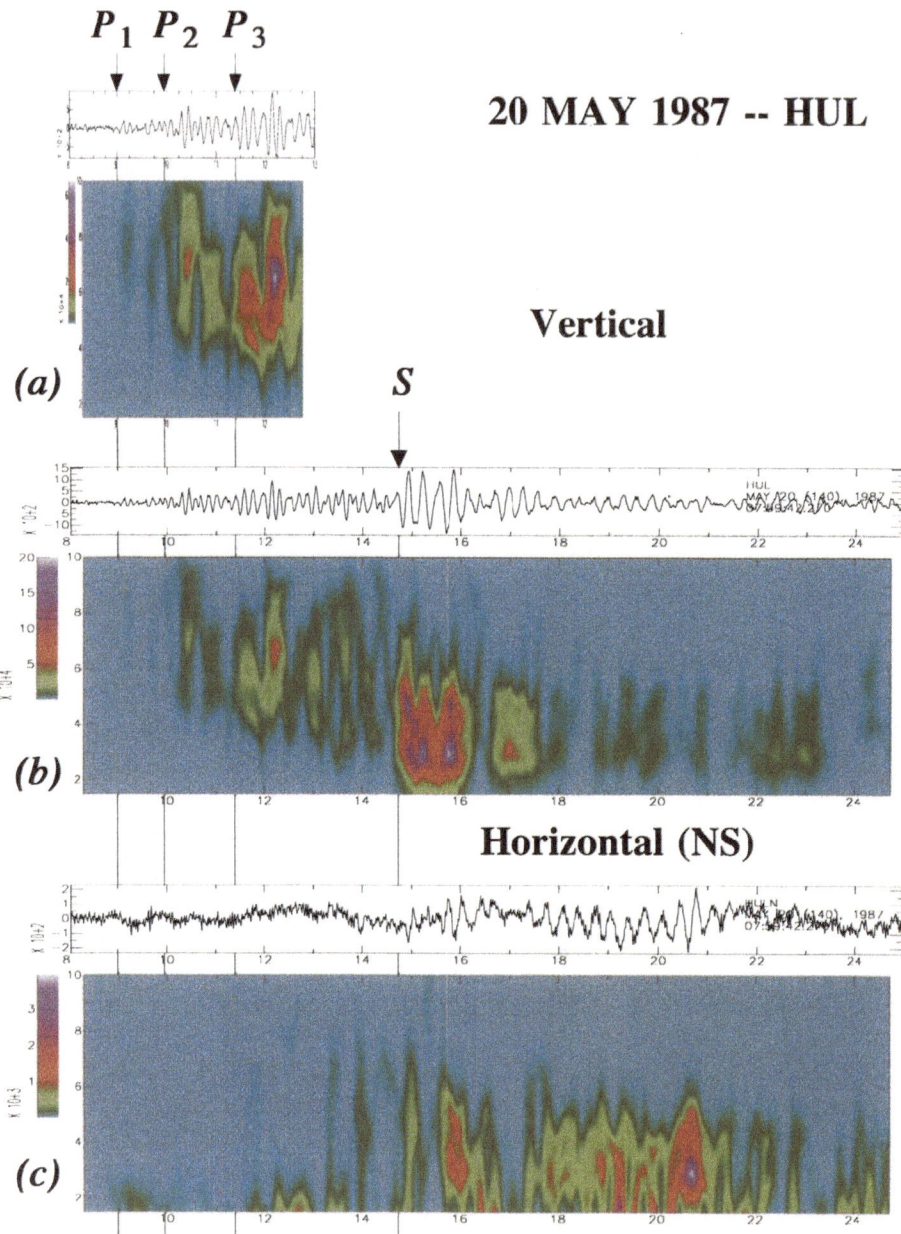

Figure 3

Spectrograms of the T phase recorded at HUL. *(a)* and *(b)*: Vertical component; *(c)*: Horizontal (north-south) component. *(a)* is a close-up of *(b)* detailing the frequency content of the $T \rightarrow P$ arrivals. All times are in seconds after trigger time. The spectrograms use a moving window of 2-s duration, sliding in increments of 0.2 s.

HUL -- 20 MAY 1987

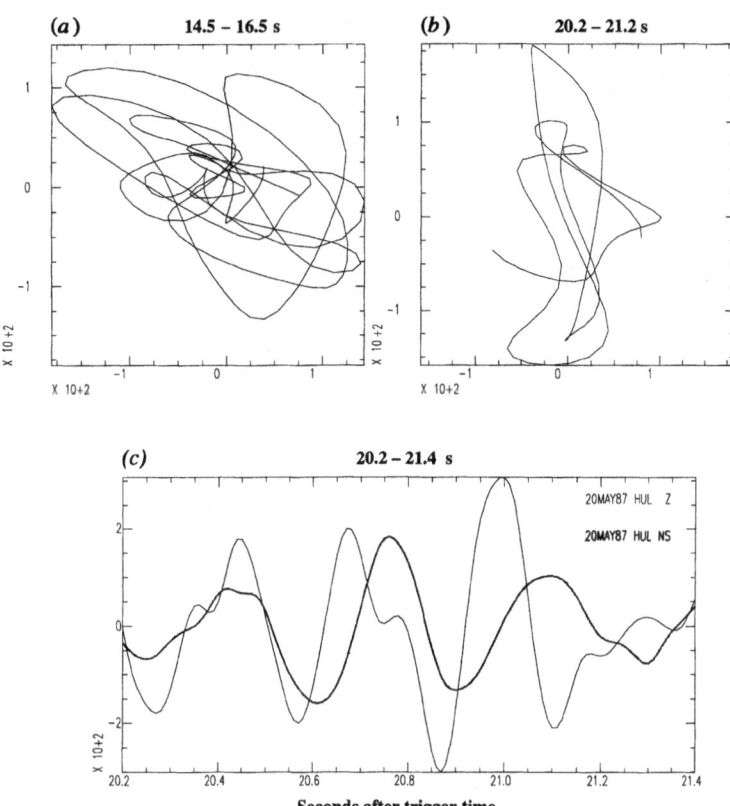

Figure 4

(a) and *(b)*: Particle motion in the horizontal plane for two windows of the later arrivals in the *T* wave. Note that the $T \rightarrow S$ conversion appears polarized in the N150°E direction *(a)*, while the later arrival *(b)* is polarized north-south. *(c)*: Comparison of the two components of the later arrival. The thin dark trace is the vertical one; the thicker, gray trace is the north-south one. Note that they are in quadrature of phase, a property characteristic of Rayleigh or interface waves (All these records have been band-pass filtered for $1 \leq f \leq 10$ Hz.).

generation of Rayleigh waves upon conversion was demonstrated theoretically by PISERCHIA *et al.* (1998).

4. Evolution across the Network

4.1. East Rift Stations

In this section, we present preliminary data regarding the evolution of the converted waves across the HVO network. We first concentrate on four stations straddling the central portion of the East Rift of Kilauea: WHA only a few hundred meters from the coastline, HUL (7 km away), KLC to the west of HUL along the rift zone, and 9 km away from the coastline, and MTV, 18 km inland. Figure 5 regroups spectrograms of the vertical components of ground motion at the four stations, for a common 20-s time window starting 5 s after trigger time.

As expected, Station WHA on the coast features a much shorter wave train than HUL. In addition, between WHA and HUL, the high-frequency components of the signal (5–8 Hz) are strongly attenuated, and the later arrival, interpreted as the $T \rightarrow S$ conversion becomes prominent at HUL. This reflects both the partition of the $T \rightarrow P$ energy into the several branches corresponding to the various crustal layers, and the resulting development of a shadow zone, as demonstrated in Paper I.

The situation at KLC is more complex; while the lower-frequency part of the wave train remains prominent, it is dispersed over several seconds. This suggests

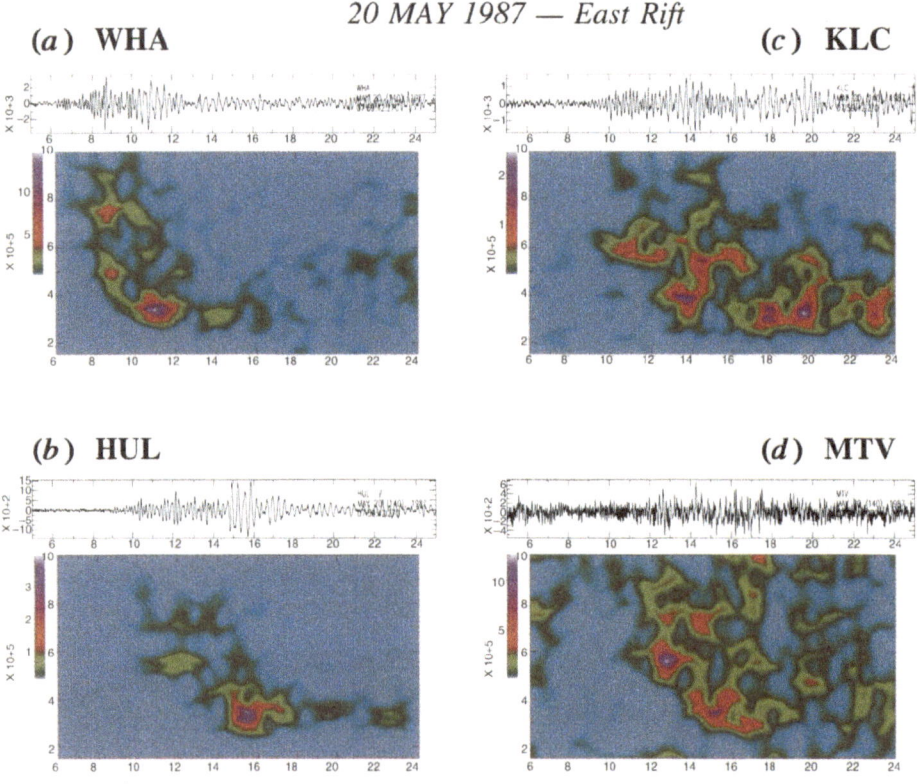

Figure 5

Comparison of spectrograms at the four East Rift stations for the event of 20 May 1987. All four windows sample between 5 and 25 s after trigger time.

multipathing to KLC, with conversions taking place at several points along the coastline. Note also that the $T \rightarrow P$ energy around 6 Hz is not damped as efficiently as it is at HUL.

Finally, when moving further inland to Station MTV, the most energetic part of the signal returns to the high-frequency (5.5 Hz) $T \rightarrow P$ conversion; this high-frequency wave train arrives only marginally later than at HUL, indicating refraction in a deeper, much faster, and probably less attenuating medium, which could be Layer 4 of the model of WARD and GREGERSEN (1973). On the other hand, the prominence of energy at high frequencies illustrates the strong damping of the later, lower-frequency, portion of the seismogram, supporting the interpretation that it corresponds to surface waves, propagating slowly across the shallow, presumably strongly fragmented and thus strongly attenuating, structures of the East Rift.

4.2. Kapoho District

We similarly study the case of the four stations located in the Kapoho District, at the easternmost end of the Big Island (Fig. 6). As in the case of WHA, the wave train at Station POI on the coast, is very short (no more than 5 s), but the $T \rightarrow S$ wave train is comparatively less developed, resulting in the strongest energy being carried by the third $T \rightarrow P$ wave packet. The wave trains at the East Rift Station KPO is perhaps most similar to HUL, with a group of high-frequency (6–7 Hz) $T \rightarrow P$ packets followed 3 s later by a low-frequency (2.5–3 Hz) $T \rightarrow S$ wave train. At KLU, located only 8 km from HUL along the East rift, the $T \rightarrow P$ wave packets are somewhat lower frequency (5 Hz rather than 7 Hz at HUL), while the $T \rightarrow S$ wave trains remain less attenuated and richer in higher frequencies (5 Hz). Finally, at HAB, beyond the rift, the distance to the coast (12 km) is insufficient to generate the fast, high-frequency phase previously observed at MTV, and the seismogram remains similar to that at KLU.

4.3. The Halape Pali-Ka'u Desert Area

We study here the conversion and propagation of the T phase in the region of the Halape Bight, located due south of Kilauea, using Stations PLO (2 km from the coast), HLP (4 km inland) and DES in the Ka'u Desert, 11 km inland (Fig. 7). With respect to other coastal stations, PLO features strongly attenuated $T \rightarrow P$ packets, with their energy remaining under 5 Hz. Similarly, HLP has all the characteristics of a shadow zone location (comparable to KLC), while DES features a revival of the $T \rightarrow P$ energy along the general lines of the MTV record, as shown in the close-up (Fig. 7*d*). The evolution of the seismogram across this area is generally similar to that along the East Rift, except for shorter distances to the coastline. This is easily explained by noting that the 1200-m isobath, along which the conversion is expected to take place, does not follow the bight, but rather develops a 10-km wide shelf, with the net result of an increase in the path of the various converted (seismic) waves.

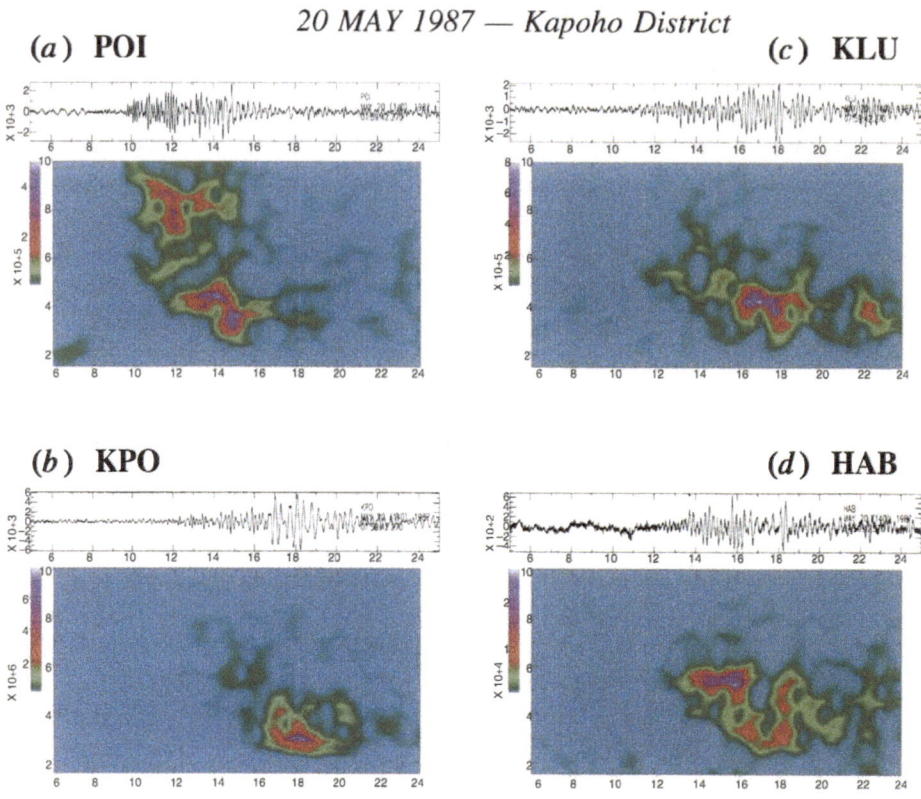

20 MAY 1987 — Kapoho District

(a) POI   (c) KLU

(b) KPO   (d) HAB

Figure 6

Comparison of spectrograms at the four stations of the Kapoho District for the event of 20 May 1987. All four windows sample between 5 and 25 s after trigger time.

4.4. Keanakolu (KKU)

This station, located on the northeastern flank of Mauna Kea, 76 km from the receiving shoreline, recorded a remarkable T wave train (Fig. 8). Despite the presence of background noise, spectrogram techniques easily pull out (i) a weak, high-frequency signal (5.5 Hz), 20.5 s after the trigger time; (ii) two lower-frequency (3.5 Hz) packets 25 and 28 s after trigger time, and (iii) another group at 42 and 46 s with comparable frequency content. In the absence of intermediate stations, it is difficult to identify the first arrival beyond doubt, but the travel-time difference between MTV and KKU would fit propagation in WARD and GREGERSEN's (1973) Layer 4, at 7.7 km/s.

We failed to identify the T phase at other distant stations of the HVO network, including Kohala (KOH; 114 km from the coast), Waikii (WAI; 75 km), Hualala'i (HUA; 71 km), and Hale Pohaku (HPU; 64 km). The recording at KKU may be

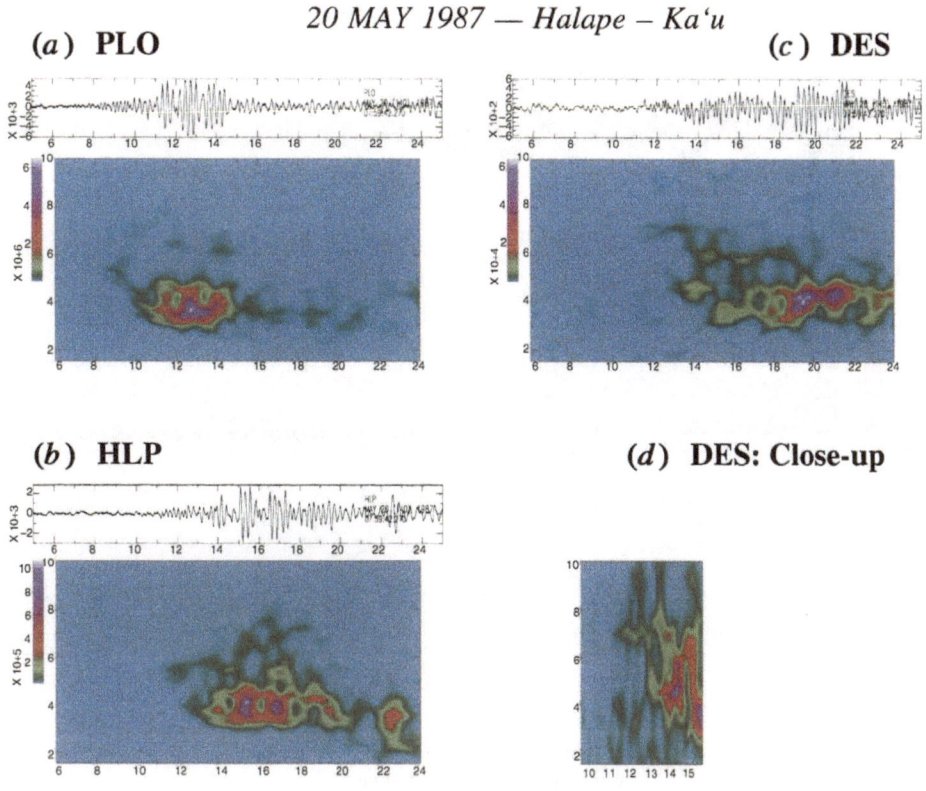

Figure 7
Comparison of spectrograms at the three stations of the Halape–Ka'u District for the event of 20 May 1987. *(a)*, *(b)*, and *(c)* sample between 5 and 25 s after trigger time. Frame *(d)* is a close-up of the seismogram at DES, similarly aligned in time, but emphasizing the re-emergence of the $P \rightarrow T$ conversion at that station, between 8 and 15 s after trigger time.

explained by the preferential location of the station, which high-frequency rays from the coast can reach by diving under the largely horizontal East Rift volcanic system, whereas seismic rays propagating to the other stations are bound to penetrate the deeply extending vertical magmatic systems under Kilauea and Mauna Loa.

5. Other Events

We briefly address here the question of the robustness of the wave shapes of *T* phases recorded at HUL from several sources at Mururoa. Figure 9 shows triggered records from all 12 events listed in Table 1, all plotted on a common vertical scale. The horizontal axes have been lagged to align all origin times (assumed to be on the

20 MAY 1987 — Keanakolu **(KKU)**

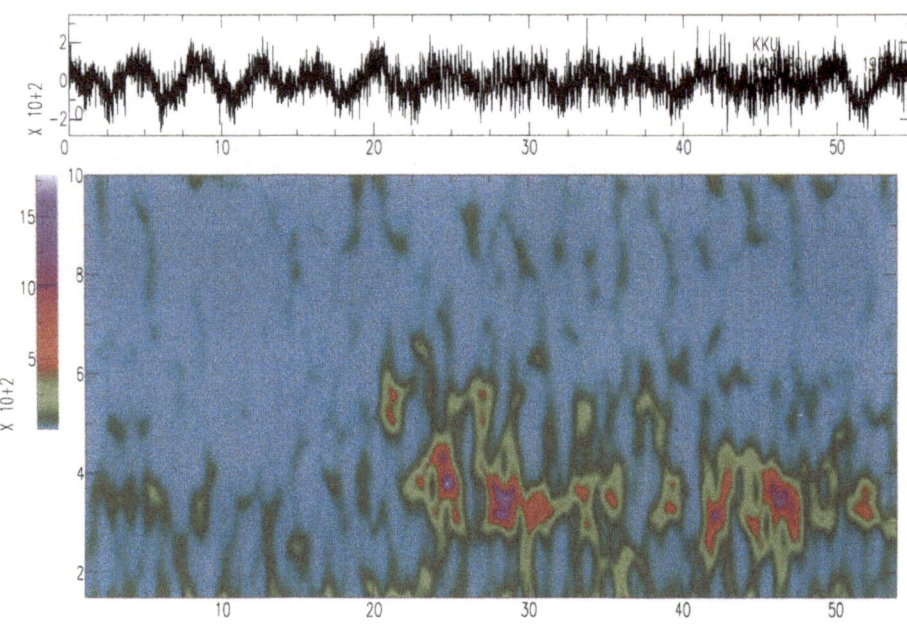

Figure 8
Spectrogram of the T phase at Keanakolu (KKU), 76 km from the shoreline. This window presents the whole triggered record, lasting 55 s.

exact minute). Several features are apparent on this figure. First, there is a confirmed relationship between the amplitude of the T wave and the size of the test, as measured by its reported m_b. While the correlation is mediocre (60%), the strongest T waves were observed from the events with the largest m_b (Fig. 10a). A similar correlation is found with the number of stations reporting to the NEIC (Fig. 10b). Concentrating then on the four events with the strongest signals (20 May 1987, 15 Jul. 1991, 25 May 1988 and 23 Nov. 1988), we find that the wave shapes of the first three are generally similar, together with their spectral contents (Fig. 11): the signals consist of a series of high-frequency (6–7 Hz) arrivals, that we interpreted as $T \rightarrow P$ conversions, followed by a lower-frequency, but generally higher amplitude $T \rightarrow S$ conversion. The fourth event, however, shows a more "blended" spectrum, in which the later arrival remains of comparatively high frequency (4–7 Hz). The origin of this feature, which is also present at WHA, is currently unclear.

Finally, the arrival times at HUL of the T phases from the 12 tests are shifted within a window of 8 s duration, which most probably reflects the precise location of the source inside the lagoon at Mururoa. In this respect, the NEIC locations are of little if any help, since many of them would plot outside the atoll.

Heiheiahulu – (HUL) — 12 Mururoa events

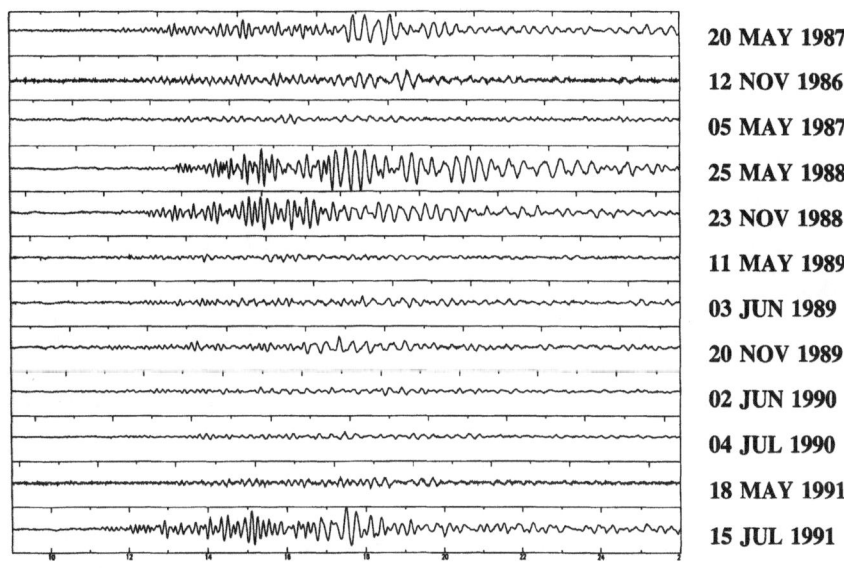

20 MAY 1987

12 NOV 1986

05 MAY 1987

25 MAY 1988

23 NOV 1988

11 MAY 1989

03 JUN 1989

20 NOV 1989

02 JUN 1990

04 JUL 1990

18 MAY 1991

15 JUL 1991

Figure 9

Comparison of the *T* wave trains from 12 Mururoa events (1986–1991) at Heiheiahulu. A common vertical
scale is used, and the seismograms have been lagged in time to adopt a common origin time, under the
assumption of a source detonated at the exact minute.

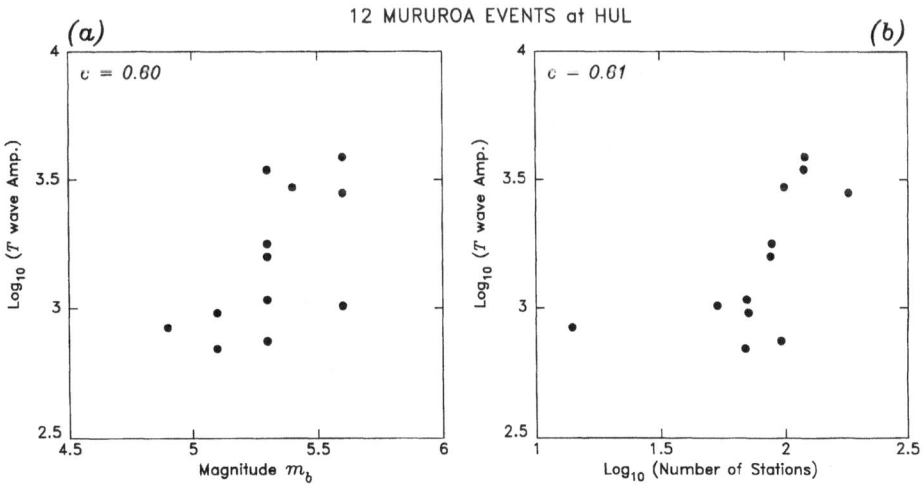

Figure 10

Peak-to-peak amplitude of the *T* wave trains recorded at HUL for the 12 Mururoa events listed in Table 1,
plotted as a function of magnitude m_b *(a)* and number of stations reporting to the NEIC *(b)*. The numbers
c at upper left are the correlation coefficients for the various datasets. Amplitude units are digital recording
units at HUL.

Heiheiahulu (HUL) — 4 Mururoa events

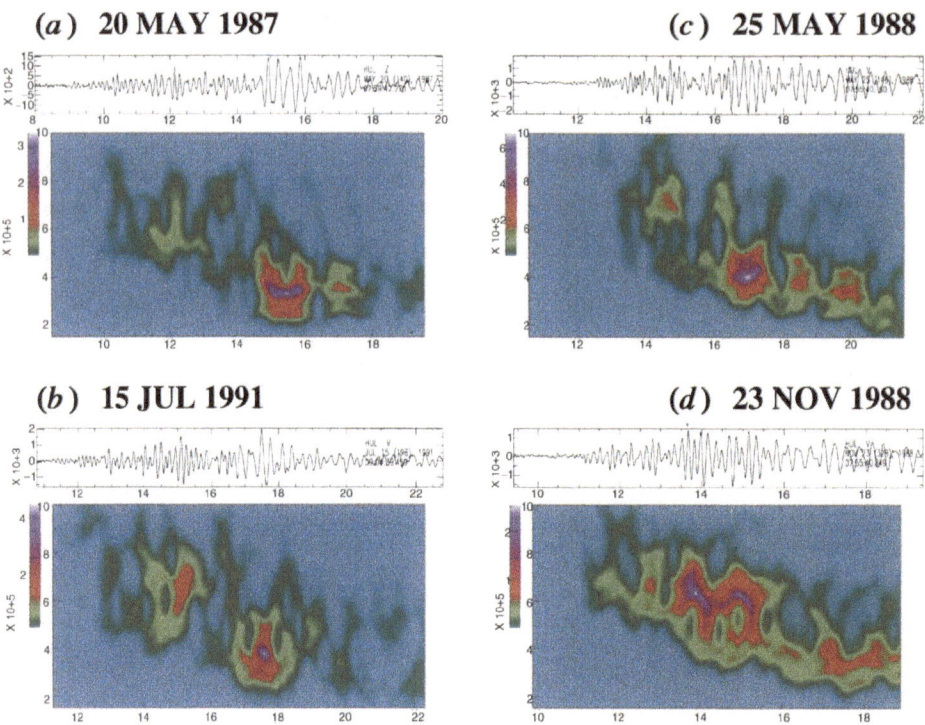

(a) 20 MAY 1987

(c) 25 MAY 1988

(b) 15 JUL 1991

(d) 23 NOV 1988

Figure 11

Comparison of spectrograms at HUL for the four largest Mururoa events. The time windows processed are 12 s long, and are lagged to align origin times (assumed to be on the exact minute). See text for interpretation.

6. Conclusions

We have examined a preliminary dataset of T phases from nuclear tests at Mururoa, recorded by the HVO network, primarily on the southern flank of the Big Island. The principal conclusions of this study are:

1. The example of the HUL record of the event of 20 May 1987 shows that the T wave is converted into a number of seismic waves featuring distinct and identifiable characteristics. At HUL, we observe separate packets of P and S energy, as well as a later, low-frequency, slow-propagating phase, interpreted as an interface, Rayleigh-type wave.

2. Propagation of the various seismic packets through the surficial structures of the island results in rapid attenuation of the high frequencies initially present at the conversion point. An estimate of the shear quality factor, $Q_\mu = 19$, is in line with the low values found in volcanic environments.

3. As the receiving site moves further inland, the characteristics of the various wave packets evolve rapidly. Most remarkable is the development of a shadow zone for the $T \rightarrow P$ conversion(s); at typical distances of 7 to 12 km, the seismogram is dominated by the $T \rightarrow S$ conversion, and consequently features a lower frequency spectrum. At larger distances, deeper refraction of P waves below the East Rift magmatic system becomes possible, with high frequencies re-emerging at MTV (18 km), and propagating efficiently as far as KKU (76 km from the shore).

4. The local crustal structure plays an important role in the exact evolution of the wave trains with distance away from the shoreline. While the general patterns are robust, the presence of a large shelf offshore can affect the position of the P shadow zone; furthermore, the magmatic systems under the Kilauea and Mauna Loa calderas essentially prevent the propagation of T phases to large distances across the island.

5. Results at HUL are generally robust for other Mururoa tests. Unknown differences in the location and condition of the shots under Mururoa Lagoon are probably the cause of arrival-time shifts of up to 8 s, and of moderate differences in the spectrum of the acoustic wave in the ocean, eventually resulting in wave packets ($T \rightarrow P$, $T \rightarrow S$) with differing frequency contents following propagation and attenuation of the converted wave in the island structure. The case of the event of 23 November 1988, featuring later arrivals with higher frequency content, remains intriguing, and will be the subject of further research.

Acknowledgments

I am grateful to Jacques Talandier for many years of stimulating collaboration in the field of Pacific T waves, and to Paul Okubo for access to the HVO archives. The maps on Figure 1 were drawn using the GMT software by WESSEL and SMITH (1991). I thank Donna Blackman for constructive comments. This research is supported by the Defense Threat Reduction Agency of the Department of Defense, under Grant DSWA01-98-1-0007.

REFERENCES

AKI, K., FEHLER, M. C., and DAS, S. (1977), *Source Mechanism of Volcanic Tremor: Fluid-driven Crack Models and their Application to the 1963 Kilauea Eruption*, J. Volcanol. Geotherm. Res. 2, 259–287.

ANONYMOUS, *The Volcano Letter*, 268 (Hawaii Volcano Observatory 1930.) pp. 1–4.

CANSI, Y., and BETHOUX, N. (1985), *T Waves with Long Inland Paths: Synthetic Seismograms*, J. Geophys. Res. 90, 5459–5465.

COOK, R. W., and STEVENS, J. L. (1998), *TP-phase Observations at the pIDC*, EOS, Trans. Amer. Geophys. Un. 79, (45), F558, (abstract).

GUILLE, G., GOUTIÉRE, G., and SORNEIN, J.-F., *Les atolls de Mururoa et de Fangataufa (Polynésie Française)*. *I. Géologie – Pétrologie – Hydrogéologie*, Commissariat à l'Energie Atomique, 168 pp., Paris, 1993.

HILL, D. P. (1969), *Crustal Structure of the Island of Hawaii from Seismic Refraction Measurements*, Bull. Seismol. Soc. Am. *59*, 101–130.

HILL, D. P., and ZUCCA, J. J. (1987), *Geophysical Constraints on the Structure of Kilauea and Mauna Loa Volcanoes and Some Implications for Seismomagmatic Processes*, U.S. Geol. Surv. Prof. Paper *1350*, 903–917.

KOYANAGI, S., AKI, K., BISWAS, N., and MAYEDA, K. (1995), *Inferred Attenuation from Site Effect-corrected T Phases Recorded on the Island of Hawaii*, Pure appl. geophys. *144*, 1–17.

LINEHAN, D. S. J. (1940), *Earthquakes in the West Indian Region*, Trans. Am. Geophys. Un. *21*, 229–232.

MCLAUGHLIN, K. L. (1997), *T-phase Observations at San Nicolas Island, California*, Seismol. Res. Letts. *68*, 296. (abstract).

OKAL, E. A., and TALANDIER, J. (1997), *T Waves from the Great 1994 Bolivian Deep Earthquake in Relation to Channeling of S-wave Energy up the Slab*, J. Geophys. Res. *102*, 27,421–27,437.

PASYANOS, M. E., and ROMANOWICZ, B. A. (1997), *Observation of T Phases across Northern California Using the Berkeley Digital Seismic Network*, EOS, Trans. Amer. Geophys. Un. *78*(46), F461–F462. (abstract).

PEKERIS, C. L. (1948), *Theory of Propagation of Explosive Sound in Shallow Water*, Geol. Soc. Am. Mem. *27*(2), 1–117.

PISERCHIA, P.-F., VIRIEUX, J., RODRIGUES, D., GAFFET, S., and TALANDIER, J. (1998), *Hybrid Numerical Modeling of T-wave Propagation: Application to the Midplate Experiment*, Geophys. J. Intl. *133*, 789–800.

RAVET, J., *Remarques sur quelques enregistrements d'ondes à très courte période au cours de tremblements de terre lointains à l'Observatoire du Faiere, Papeete, Tahiti*, Sixth Pacific Sci. Congress, vol. 1, 1940, pp. 127–130.

SMITH, J. R., *Island of Hawaii and Loihi submarine volcano, high-resolution multibeam bathymetry around the Island of Hawaii [1:75,000, 1:250,000, 1:500,000]*, Sheet 6, Hawaii Seafloor Atlas, Hawaii Institute of Geophysics and Planetology, Honolulu, 1994.

TALANDIER, J., and OKAL, E. A. (1979), *Human Perception of T Waves: The June 22, 1977 Tonga Earthquake Felt on Tahiti*, Bull. Seismol. Soc. Am. *69*, 1475–1486.

TALANDIER, J., and OKAL, E. A. (1996), *Monochromatic T Waves from Underwater Volcanoes in the Pacific Ocean: Ringing Witnesses to Geyser Processes?*, Bull. Seismol. Soc. Am. *86*, 1529–1544.

TALANDIER, J., and OKAL, E. A. (1998), *On the Mechanism of Conversion of Seismic Waves to and from T Waves in the Vicinity of Island Shores*, Bull. Seismol. Soc. Am. *88*, 621–632.

WARD, P. L., and GREGERSEN, S. (1973), *Comparison of Earthquake Locations Determined with Data from a Network of Stations and Small Tripartite Arrays on Kilauea Volcano, Hawaii*, Bull. Seismol. Soc. Am. *63*, 679–711.

WESSEL, P., and SMITH, W. H. F. (1991), *Free Software Helps Map and Display Data*, EOS, Trans. Am. Un. *72*, 441 and 445–446.

(Received April 30, 1999, revised August 19, 1999, accepted September 2, 1999)

To access this journal online:
http://www.birkhauser.ch

Pure appl. geophys. 158 (2001) 475–512
0033–4553/01/030475–38 $ 1.50 + 0.20/0

© Birkhäuser Verlag, Basel, 2001

❚ Pure and Applied Geophysics

Normal Mode Composition of Earthquake T Phases

GERALD L. D'SPAIN,[1] LEWIS P. BERGER,[1] W. A. KUPERMAN,[1]
JEFFRY L. STEVENS,[2] and G. ELI BAKER[2]

Abstract — Understanding the nature of the coupling between the underwater acoustic field and the land seismic field is important for evaluating the performance of the T-phase stations in the International Monitoring System for the Comprehensive Nuclear-Test-Ban Treaty. For upslope propagation in an ocean environment, the places where underwater acoustic field energy couples into the land seismic field are determined to first approximation by the local water depth and the normal mode composition of the acoustic energy. Therefore, the use of earthquake-generated T phases as natural probes of water-to-land coupling characteristics is aided by knowledge of their modal composition. Data collected by a 200-element, 3000-m-aperture vertical hydrophone array during a 1989 experiment in the deep northeast Pacific Ocean are used to determine the mode composition of T-phase arrivals from two m_b 4.1 earthquakes near the west coast of the U.S., one occurring offshore and the other on land. Results from an eigenanalysis approach and conventional mode decomposition for the two events are consistent and show that at 5 Hz, the offshore event's arrivals have higher-order mode content compared to those from the event on land. Single hydrophone recordings at Pt. Sur of two m_b 4.4 Hawaiian events in 1996 and 1997, one occurring offshore and the second on land, display time-frequency arrival structures that are explainable by the dispersion characteristics over the oceanic path. Although other effects due to complex source time functions and shear wave and dispersive propagation effects along the initial land path cannot be separated with these single element data, differences in these two events' arrival structures suggest differences in normal mode content consistent with those seen in the pair of 1989 events. Ocean-path dispersion also appears to play a significant role in determining the in-water arrival structure from a 1995 French nuclear test at Mururoa. Recordings of two Hawaiian events in 1997 by the T-phase station VIB and the seismic station at Berkeley illustrate that the water-land coupling confuses the relative timing between normal modes, resulting in apparent loss of information about the source.

Key words: CTBT, IMS, T phase, hydroacoustic, normal modes.

Introduction

The hydroacoustic network of the International Monitoring System (IMS) for the Comprehensive Nuclear-Test-Ban Treaty (CTBT) is a set of globally distributed stations each composed of either at most a few underwater hydrophones or a land-based seismic sensor system, referred to as a "T-phase station." The network ultimately will be comprised of only five hydroacoustic stations and six T-phase

[1] Marine Physical Laboratory, Scripps Institution of Oceanography, San Diego, CA 92152, USA.
[2] Maxwell Technologies, San Diego, CA 92123-1506, USA.

stations (COMPREHENSIVE NUCLEAR-TEST-BAN TREATY ORGANIZATION, 1998). Because of the sparse station coverage and the limited number of sensor elements at each station, as much information as possible must be extracted from single sensor data. In addition, understanding the mechanisms of underwater sound coupling into the land seismic field is important for interpreting the data from, and evaluating the performance of, the *T*-phase stations.

One way to evaluate the performance of land-based *T*-phase stations is to use naturally-occurring *T* phases as probes of opportunity to study the water-to-land coupling at the frequencies and spatial scales of relevance to the CTBT. A good data set for this purpose has been obtained from the recordings of events that occurred on the main island of Hawaii and at Loihi Seamount, the newly forming Hawaiian Island just south of the main island. Hawaiian earthquakes have long provided signals for *T*-phase studies (e.g., TOLSTOY and EWING, 1950; BYERLY and HERRICK, 1954; JOHNSON *et al.*, 1963; NORTHROP, 1974). Features of the 1996 earthquake swarm at Loihi are described in CAPLAN-AUERBACH *et al.* (1996), DUENNEBIER *et al.* (1997), HAWAII CENTER FOR VOLCANOLOGY (1998), and CAPLAN-AUERBACH and DUENNEBIER (2000). Data have been acquired from the single underwater hydrophone station at Pt. Sur and from several land stations in western North America, including those in the Berkeley seismographic network (BDSN), those in the set of central California stations operated by the Pacific Gas and Electric Company, and the IMS *T*-phase station VIB at Van Inlet. These data have been used to derive empirical estimates of the transfer function of the water-to-land coupling along the North American West Coast, and to determine the propagation mechanisms of this converted seismic energy across California (Stevens *et al.*, 1998). Numerical modeling of these broadband transfer functions is being done in both two and three spatial dimensions (i.e., 2-D, N × 2-D, and 3-D) as well as in time using a hybrid normal mode/elastic finite difference approach (Stevens *et al.*, 2000).

To develop a sense of the coupling physics and provide a guide for more computationally intensive numerical modeling, a simple model based on the results of acoustic propagation in a wedge-shaped ocean can be used. Considerable research has been conducted by the underwater acoustics community on this problem (RANGE-DEPENDENt BENCHMARK PROBLEMS in OCEAN ACOUSTICS, 1990). The classical model of upslope propagation in a penetrable wedge at a single frequency (for example, see the picture on the cover of JENSEN *et al.*, 1994) shows the water column modes dumping their energy into the bottom, forming a discrete beam for each mode in the bottom. These discrete mode beams occur in descending mode order with decreasing water depth as mode cutoff is reached. Both the numerical modeling results (JENSEN and KUPERMAN, 1980) and laboratory measurements (COPPENS and SANDERS, 1978) indicate that the modes propagate and couple into the bottom "adiabatically." That is, very little higher mode energy cascades into the lower order modes. As the frequency decreases, the range intervals at which the modes couple into the ocean bottom occur farther offshore; for increasing frequency, they occur closer to the coast. This classical model indicates where along the ocean bottom the coupling occurs, and so where the bottom properties are most important. It describes, at least qualitatively, the general features seen in many on-land *T*-phase recordings.

The reciprocal of the simple model of propagation in a wedge-shaped ocean provides one mechanism for the coupling of land seismic energy into the underwater acoustic field. In this case, the seismic energy at a given frequency couples into the underwater acoustic field through the discrete mode beams in the bottom. Therefore, the energy that couples into the higher order waterborne modes at a given frequency travels for a longer distance at the higher land seismic velocities than the lower mode energy. This point leads to the prediction that late-arriving energy in the water column from a land-based seismic event is predominantly composed of lower order modes. The character of underwater recordings from land events presented in this paper is consistent with this prediction. This reciprocal model differs from those that rely upon a rough surface scattering mechanism or on interface waves as an intermediate propagation pathway. It only requires a way of focusing the bottom energy into the discrete mode beams. The gradients with depth of the seismic velocities play an important role in this regard (COLLINS, 1995). This model actually is similar to that of "downslope conversion" (MILNE, 1959; GRINDA, 1960; JOHNSON *et al.*, 1963) i.e., *T*-phase generation through repeated reflection from the ocean surface and bottom in a bottom-downsloping environment. However, with the full wavefield approach here, the coupling occurs in a way such that subsequent interactions with the bottom occur at angles past the critical angle so that no waterborne energy is retransmitted into the bottom. This reciprocal coupling model has provided the framework for the analysis of offshore underwater acoustic recordings of on-land vehicle activity (D'SPAIN *et al.*, 2001).

Naturally-occurring *T* phases are used in this study as probes of the water-land coupling characteristics. Their use for this purpose is greatly aided by an understanding of their normal mode composition in the water column. In Section I, underwater *T*-phase recordings made by the Long Vertical Line Array (LVLA) are analyzed using an eigenanalysis approach as well as conventional mode decomposition to determine the relative contributions of the normal mode components. The focus is on the recordings of two earthquakes, both having the same body wave magnitude of 4.1 and similar epicentral distance from the LVLA. However, one event occurred offshore in approximately 3000-m-deep water and the other occurred on land, leading to differences in their deep water arrival structure and normal mode content.

Data from receiving systems having vertical aperture generally are not available. However, in some cases, a rough estimate of the normal mode composition may be obtainable from recordings made by a single hydrophone. This possibility is discussed in Section II, where the arrival structure at the Pt. Sur station from natural events in Hawaii and Loihi in 1996 and 1997, and the man-made detonations in French Polynesia in 1995, are examined. As shown in this section, the dispersion characteristics of low frequency, underwater sound propagation have an observable influence on the arrival structure of these events. In fact, the dispersed nature of in-water recordings of underwater, short-duration events, like underwater explosions, at long range contains useful information such as the range to the source (EWING and WORZEL, 1948; KUPERMAN *et al.*, 2001). However, once these signals strongly interact with the ocean bottom and couple into land vibrations, this information appears to be lost and the arrival structure on land then is determined mainly by the character of the water/land coupling.

Finally, Section III presents a summary of the conclusions from this work.

I. Mode Composition of T Phases in Deep Water

Knowledge of the normal mode composition of ocean-borne acoustic signals is important to understanding how and where these sounds couple into land-based vibrations. A determination of the mode composition is most easily accomplished with data from sensor systems which have vertical aperture spanning a significant portion of the water column. Such data were collected during the 1989 VAST (Various Arrays Sea Test) experiment conducted in the northeast Pacific Ocean. The purpose of this section is to use VAST data to determine the mode composition of naturally-occurring *T* phases in the deep ocean.

A. LVLA Recordings of T Phases during the VAST Experiment

For the VAST experiment, the Marine Physical Lab. (MPL) designed, built, and deployed a 3000-m aperture vertical line array of hydrophones (LVLA) from R/P FLIP (FLoating Instrument Platform). Throughout its history, FLIP, a 100-m-long manned spar buoy, has provided a stable platform from which *T*-phase measure-ments have been made (NORTHROP and JOHNSON, 1965). The data from each of the 200 equally-spaced array elements were digitized at a 250 samples/s rate and recorded on FLIP over an 11-day period in July, 1989. Additional information on the LVLA hardware and the VAST experiment can be found in OLIVERA *et al.* (1994), SOTIRIN and HILDEBRAND (1988), HODGKISS (1993), D'SPAIN *et al.* (1991), and HOWE *et al.* (1991). To our knowledge, this array is the longest vertical aperture deployed to date. Thus, its data are some of the best available for examining the mode content of deep water *T* phases.

Figure 1 depicts a plot of the array element positions as a function of depth along with two sound speed profiles measured during the experiment. Only the upper 5000 m of the 5160-m-deep water column are shown. Every fourth element of the array, starting with the deepest element, is indicated in the figure. Given the low frequency content of the *T*-phase signals of interest, only the data from these elements were used in the analysis presented in this paper. Even with the large aperture, the lower two-fifths of the water column were not sampled. This lack of full

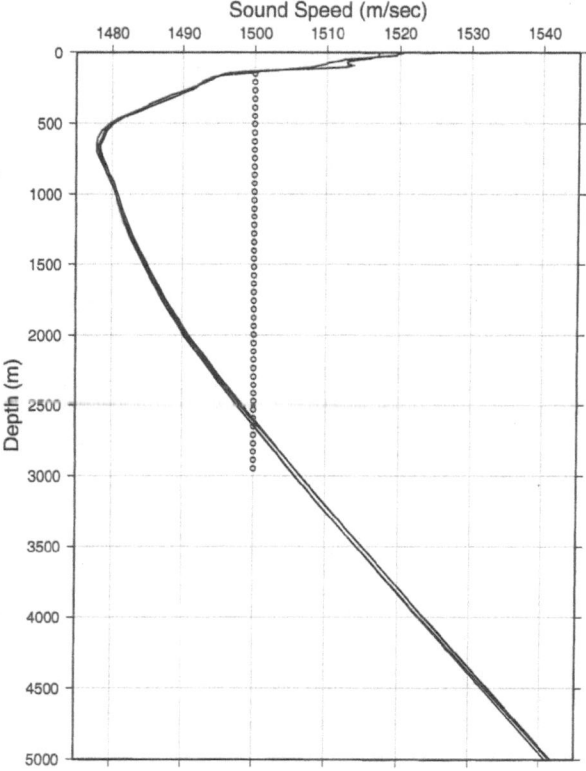

Figure 1

Two sound speed profiles derived from CTD data collected at the FLIP location during the July, 1989 VAST experiment. Also plotted are the depths of the 48 LVLA array elements (every fourth element) whose data are processed for the results presented in this paper.

water column coverage results in a lack of perfect orthogonality of the normal modes across the array aperture. As shown in the figure, the array was optimally positioned to record long-distant signals trapped in the deep sound channel.

The location of the LVLA in the northeast Pacific Ocean is shown on the map in Figure 2. For reference, the locations of the Wake Island and Pt. Sur hydroacoustic stations, the VIB T-phase station, and the Loihi seamount just south of the main island of Hawaii also are plotted on the map. The LVLA was located near the great circle path between Loihi and Pt. Sur, and nearly equidistant from these two locations.

Several *T*-phase events were recorded during the VAST experiment. The recordings of the five earthquake events in Table 1 have been fully analyzed using the techniques described here.

As indicated in Table 1, the body-wave magnitudes for events 3 and 4 are equivalent. In addition, their locations near the Pt. Sur station and their epicentral

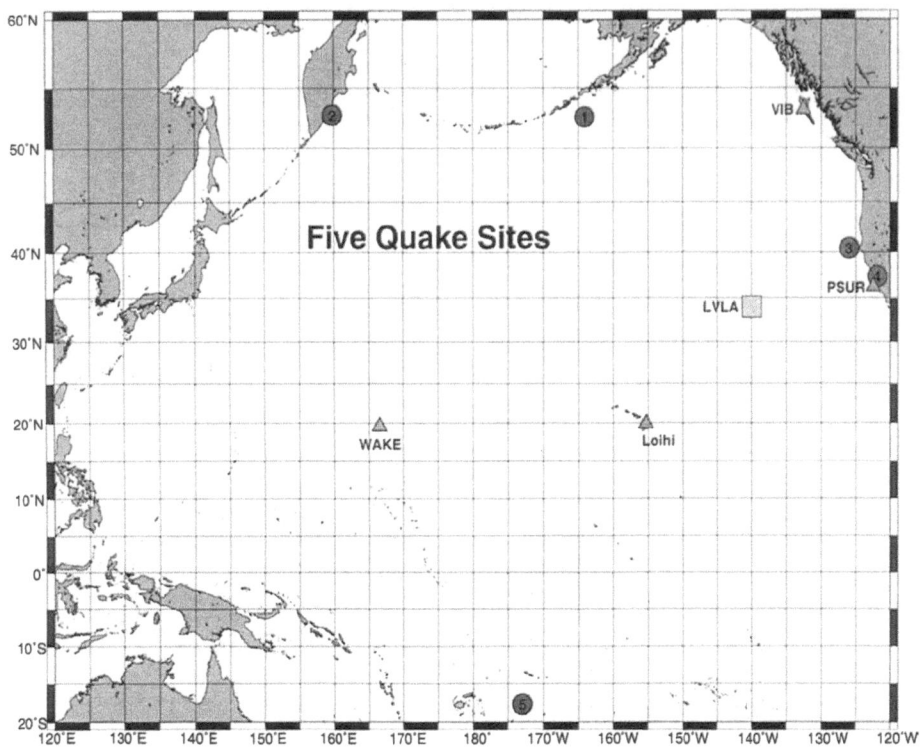

Figure 2

Map of the northern part of the Pacific Ocean showing the location of FLIP and the LVLA during the VAST experiment at 34.0°N, 140.0°W (square), along with the five earthquake events (circles numbered 1 through 5, with the numbers corresponding to those in the first column of Table 1) whose LVLA recordings have been analyzed using the techniques described in this paper. For reference, the locations of the Wake Island and Pt. Sur hydroacoustic stations, the VIB *T* phase station, and the Loihi seamount just south of the main island of Hawaii also are plotted (triangles) on the map.

Table 1

Five earthquakes analyzed in the 1989 LVLA data

Quake no.	Location	JD	Time (GMT)	Latitude	Longitude	m_b	Range (km)
1	Aleutian Isl.	189	10:56:54	52.65	−164.05	5.1	2819
2	Kamchatka	189	09:31:57	52.84	159.86	5.5	5123
3	Mendocino Ridge	194	14:41:42	40.42	−125.91	4.1	1437
4	San Jose	190	13:38:44	37.40	−121.79	4.1	1688
5	Tonga Ridge	189	00:48:09	−17.72	−172.95	5.1	6706

Epicentral locations, origin times, and body-wave magnitudes for these events were obtained from NATIONAL EARTHQUAKE INFORMATION CENTER (1998).

distances from the LVLA are similar. The major difference is that event 3 occurred offshore, under a 3000-m-deep water column, whereas event 4 occurred on land. The results for these two events are the focus of the discussion.

The calibrated single-element spectrograms for events 3 and 4 in the 0 to 25 Hz band over a 5-min. period are shown in Figures 3 and 4. The LVLA data acquisition system has a flat response to pressure from 3 Hz to 110 Hz, and the roll-off in response below the high-pass filter corner at 3 Hz was not taken into account in these

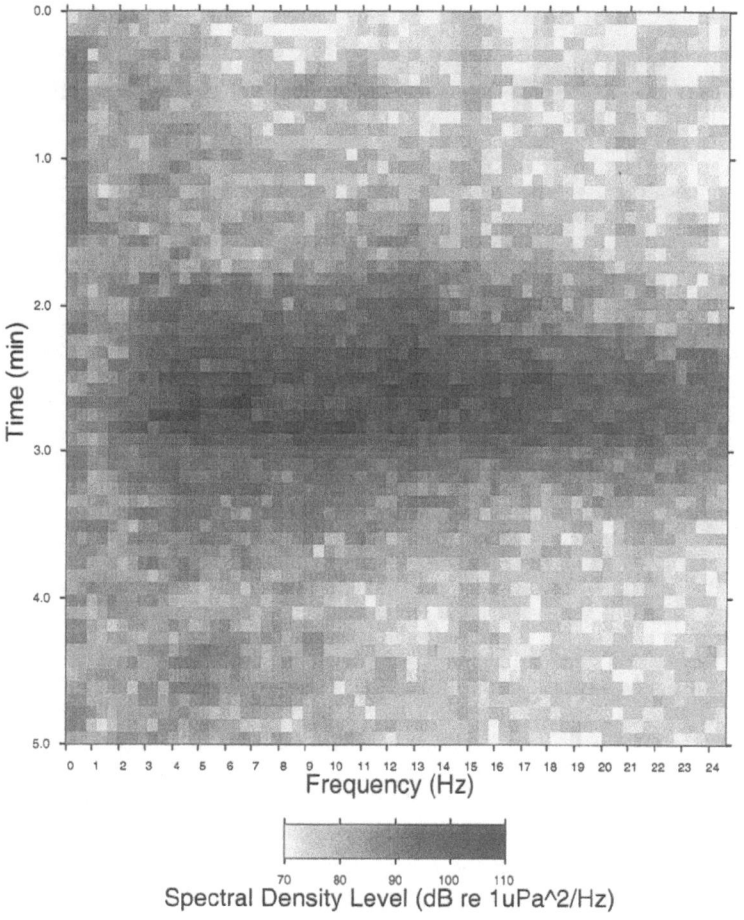

Figure 3

Calibrated single-element spectrogram from 0 to 25 Hz over a 5-min. time period encompassing the arrival of the *T* phase generated by the m_b 4.1 event at the Mendocino Ridge (event "3" in Figure 2 and in Table 1). This plot is estimated from data collected by element 5 of the LVLA array at 2708 m depth. The rolloff in system response below the high-pass filter corner at 3 Hz is not taken into account in this plot nor in Figure 4.

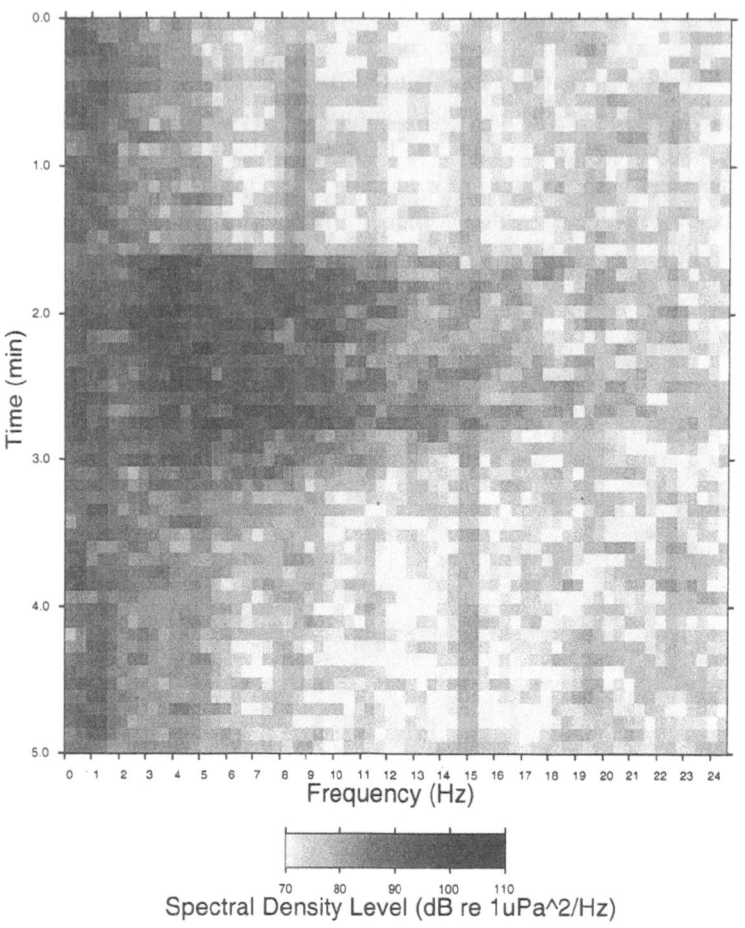

Figure 4
Calibrated single-element spectrogram for the *T*-phase arrival generated by the m_b 4.1 event on land near San Jose, CA (event "4" in Fig. 2 and in Table 1). Other aspects of this plot are the same as in Figure 3.

two figures. The element whose data are used for these plots was near the bottom of the array at a depth of 2708 m. The received signals from both events last more than a minute, however those from the offshore event are higher in level and extend to higher frequencies than those from the land-based event.

The plane wave vertical arrival structure at 5 Hz over the same 5-min. periods of time for both events is shown in Figures 5 and 6, respectively. The arriving *T*-phase energy for event 3 is greatest near angles of $\pm 10°$ whereas it is concentrated about the horizontal in event 4. Since the equivalent ray angles for a given normal mode increase from the horizontal with increasing mode number, these vertical direction-

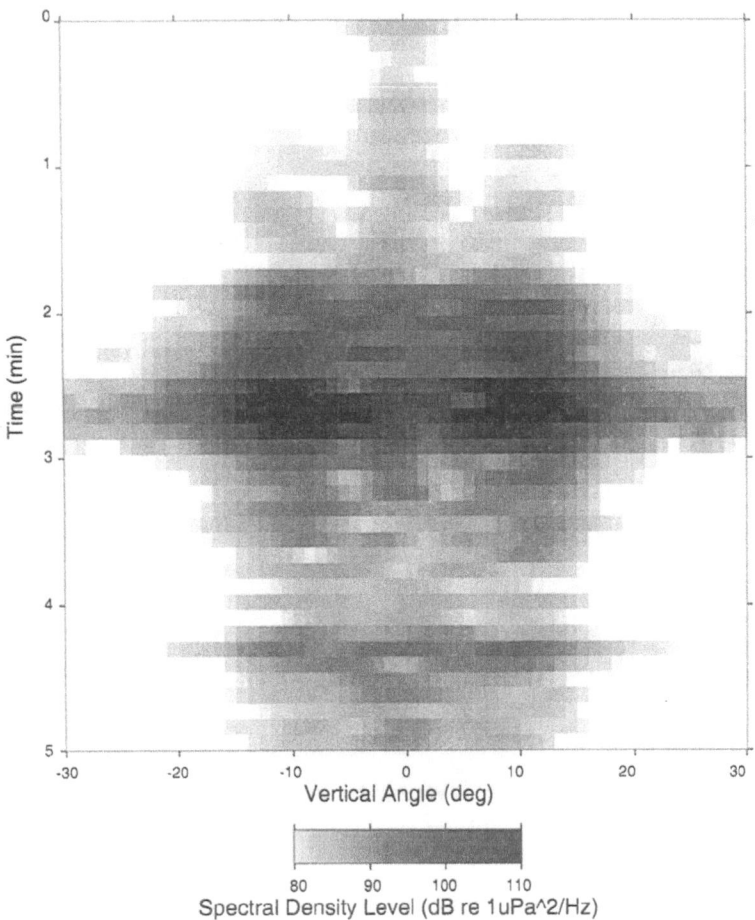

Figure 5

The vertical directionality at 5 Hz obtained from conventional plane wave beamforming of the LVLA data over the 5-min. time period encompassing the Mendocino Ridge event (i.e., the same period as in Fig. 3). The time series from every fourth element of the array were low-pass filtered and desampled to 50 Hz and then used to form consecutive, non-overlapped averaged data cross spectral matrices. These matrices were derived by averaging four outer products, obtained from 128-point Fourier transforms which have a 50 percent overlap after windowing with a Kaiser-Bessel window of α equal to 2.5. The vertical directionality at each time snapshot then was calculated by pre- and post-multiplying the averaged data cross spectral matrix by the plane wave steering vectors at 1° increments from $-30°$ (downward from the horizontal) to $+30°$ (upward from the horizontal).

ality results suggest that the offshore Mendocino Ridge earthquake generated greater higher-order mode energy at 5 Hz than the land-based San Jose earthquake. The taper to horizontal angles over the final three-quarters of a minute or so in event 4's arrival structure is at least partly caused by the characteristics of the coupling of

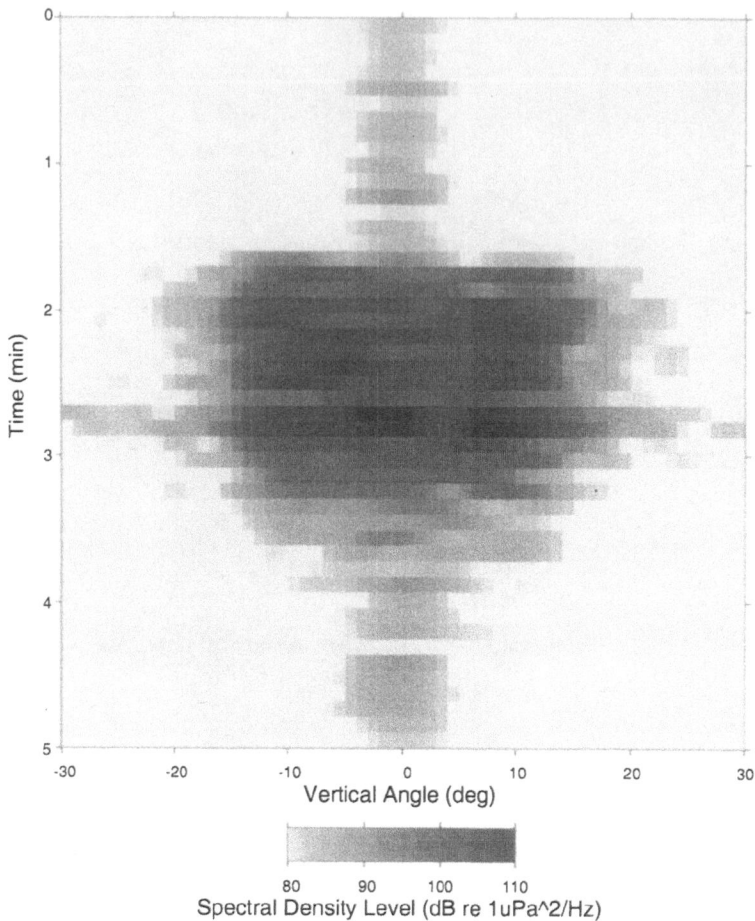

Figure 6
The vertical directionality at 5 Hz for the San Jose event (number 4) over the same 5-min. time period as in
Figure 4. All other aspects of the plot are the same as in Figure 5.

land-based seismic energy into the water column, as discussed in the Introduction. That is, this energy arriving near the horizontal represents the lowest order modes, which travel the shortest distance on land because they couple into the water column at the shallowest depths.

B. *Conventional Normal Mode Decomposition*

The solution of the acoustic wave equation at a single circular frequency, ω, due to a single point source with complex amplitude $A(\omega)$, in terms of the adiabatic normal mode approximation can be written as (e.g., JENSEN *et al.*, 1994):

$$p(r, z_s, z, \omega) = A(\omega) \sum_n a_n(\omega, r) \Psi_n^s(z_s) \Psi_n^r(z) \exp\left[i \int \kappa_n \, dr\right], \tag{1}$$

where the mode amplitudes are:

$$a_n(\omega, r) = \frac{1}{\rho(z_s)} \left[\frac{1}{8\pi \int \kappa_n \, dr}\right]^{1/2} \exp[-i(\omega t + \pi/4)], \tag{2}$$

the mode eigenfunctions at the source location and the receiver location are $\Psi_n^s(z_s)$ and $\Psi_n^r(z)$, respectively, and the integral of the mode wave number, κ_n, is over the range between source and receiver. The expression in Equation (1) is valid for weakly range-dependent media and at ranges from the source greater than several water depths.

The set of normal modes form an orthonormal set of basis functions across the full depth of the waveguide, i.e.,

$$\int_0^D \frac{1}{\rho(z)} \Psi_n^r(z) \Psi_m^r(z) \, dz = \begin{cases} 1 & \text{for } m = n \\ 0 & \text{for } m \neq n \end{cases} \tag{3}$$

where the scaling is done by the medium density, $\rho(z)$. Multiplying both sides of Equation (1) by $\rho^{-1}(z) \Psi_m^r(z)$ and integrating over the full waveguide depth gives:

$$\int_0^D \frac{1}{\rho(z)} p(r, z) \Psi_m^r(z) \, dz = A(\omega) a_m \Psi_m^s(z_s) \exp\left[i \int \kappa_m \, dr\right]. \tag{4}$$

Therefore, the inner product with depth of the received complex pressure at a given frequency with the mode eigenfunction at that frequency gives the mode amplitude. This approach has been used to determine the normal mode content of higher-frequency tones generated by towed underwater projectors during the VAST experiment (OLIVERA et al., 1994).

Normal mode eigenfunctions for the deep water environment at the VAST FLIP site were generated with the Kraken computer code (PORTER, 1991). The results of performing the inner product of the Kraken mode eigenfunctions with the complex pressure across the LVLA at 5 Hz are shown in Figures 7 and 8, for the Mendocino Ridge event and the onshore San Jose event, respectively. Clearly, the arrivals from both events have significant energy distributed across several normal modes. The energy distribution is concentrated more in the lower order modes for the onshore event (Fig. 8) than the one offshore (Fig. 7).

This point is illustrated more effectively in Figure 9 where the linear magnitude of the mode amplitudes has been averaged over the time period of the main T-phase arrival before conversion to logarithmic units. The energy is approximately evenly distributed among the first 4 modes in event 3's arrival, whereas only the lowermost 3 modes contribute appreciably to the arrival from event 4.

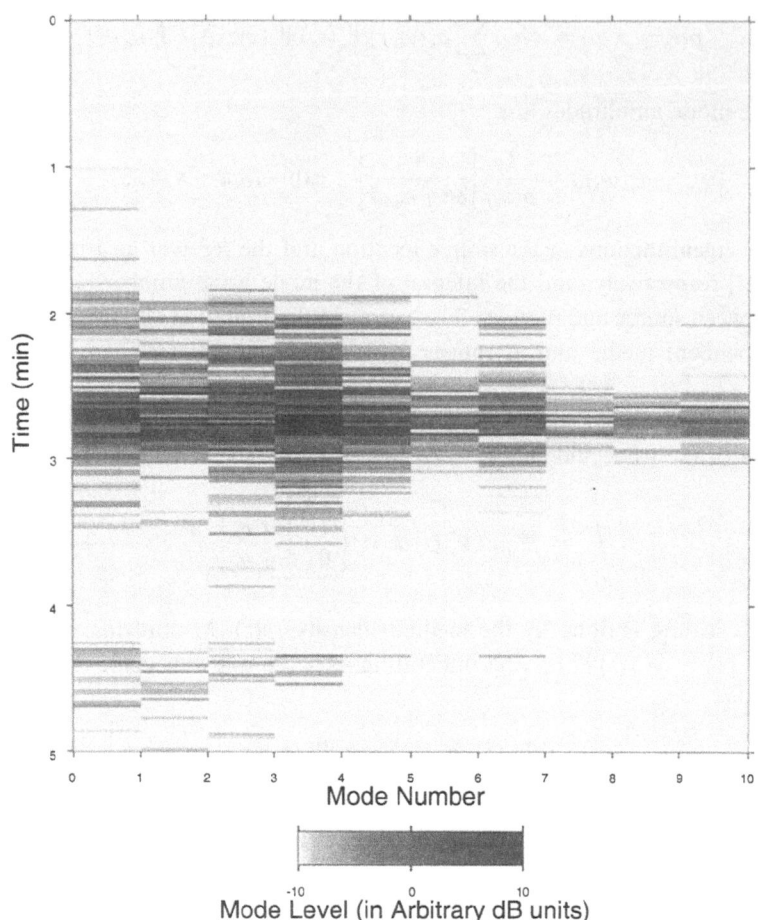

Figure 7

Conventional mode decomposition results at 5 Hz for event 3 at the Mendocino Ridge, using data from every fourth element of the LVLA array. The vertical time scale is the same 5-min. period as in Figures 3 and 5. The inner product of the complex pressure at 5 Hz across the array and the Kraken-derived eigenfunctions have been converted to arbitrarily scaled decibel units by taking twenty times the logarithm of the magnitude of the inner product, after normalizing by a constant, arbitrary magnitude value.

Figure 8 also shows the evolution to lower mode number with time over the last 3/4-min. of event 4's arrival, as suggested in the vertical directionality results.

To perform the mode decomposition just presented, a numerical code first must be used to calculate the mode eigenfunctions. Therefore, accurate knowledge of the sound speed profile, bottom geoacoustic properties, and array element positions is required. In addition, any variations in the sensitivities of the array elements must be known and taken into account when forming the inner product. Mismatch between the assumed and true environmental and array properties will result in degradation

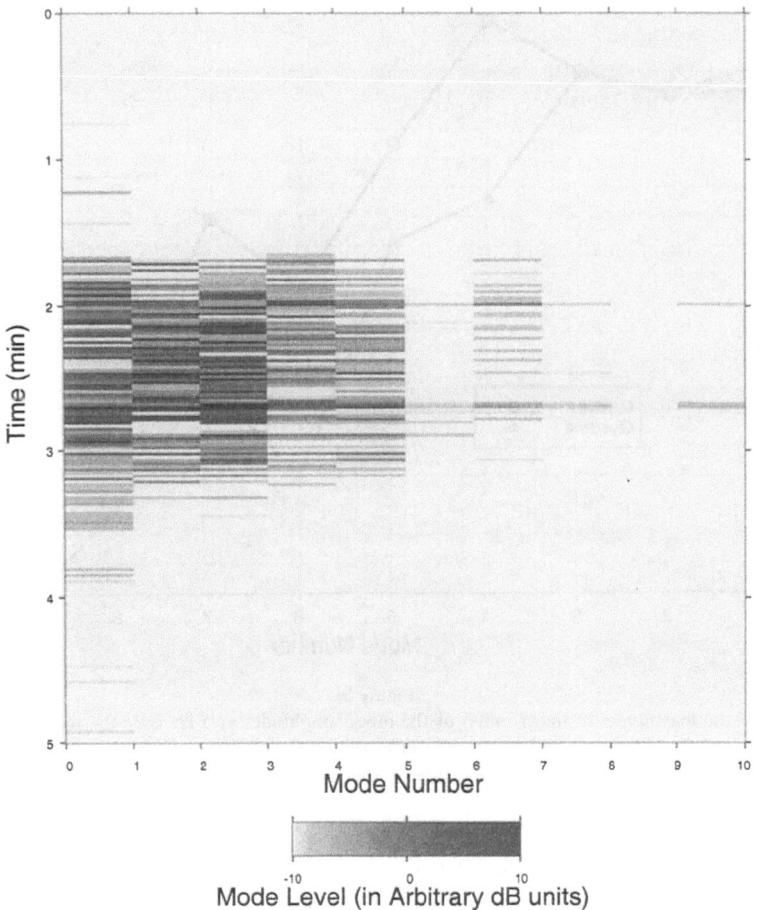

Figure 8
Conventional mode decomposition results at 5 Hz as a function of time for event 4 near San Jose. The 5-min. time period along the vertical axis is the same period as in Figures 4 and 6. The arbitrary normalizing constant has the same value as in the previous figure.

of the mode amplitude estimates. Another source of error results from the limited vertical extent of the array. Equation (3) indicates that the modes are orthonormal only over the full extent of the waveguide (which may include the sub-bottom, although modes that interact significantly with the bottom do not contribute appreciably to signals received at the large ranges of interest here). Lack of orthogonality of the modes over the LVLA aperture will result in "cross talk," i.e., leakage, between the modes. To illustrate the degree of intermode leakage, Figure 10 displays a plot of the magnitude of the inner products of the first ten modes at 5 Hz across the LVLA aperture. The leakage is minimal for the first few modes, but

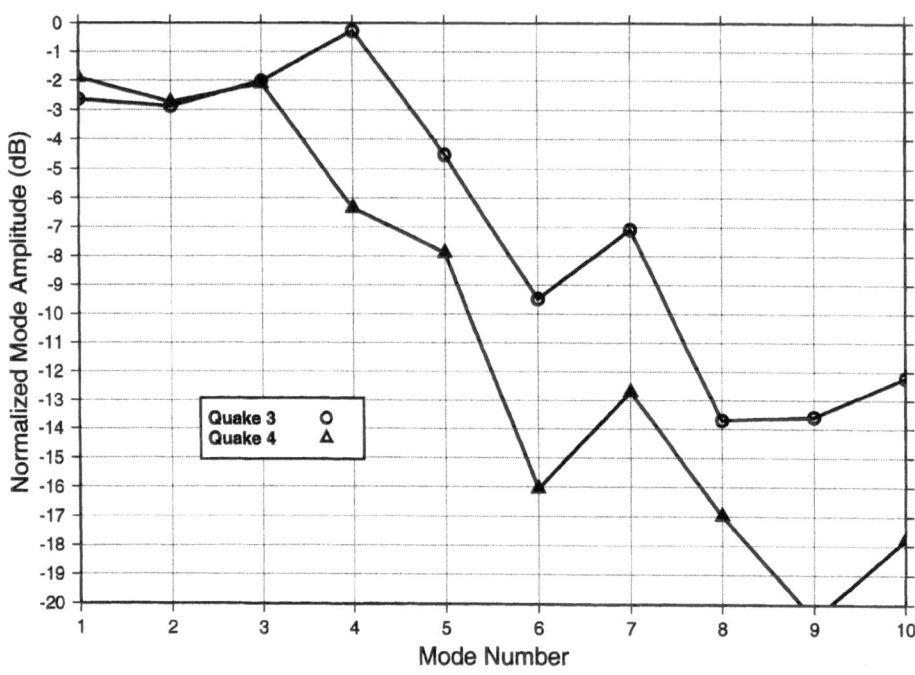

Figure 9
Average of the magnitude (in linear units) of the mode amplitudes at 5 Hz over the time period of the
T-phase arrivals from events 3 and 4. The respective time series of the mode amplitudes from which the
averages are calculated are plotted in Figures 7 and 8 over the time interval from 1.5 to 4.0 min.

becomes appreciable for the higher order modes. However, since much of the cross
talk occurs between neighboring modes, its impact on the results is not too severe.

In addition, the vertical span of the LVLA may miss a significant portion of
energy associated with a mode. This issue is particularly relevant to those modes
whose equivalent rays have lower turning points within the water column but below
the lowermost element of the array. This effect results in a decrease in the value of the
inner mode product along the diagonal in Figure 10 with increasing mode number,
reaching a minimum value of just slightly greater than 0.4 at mode 7. That is, only
slightly more than 40 percent of the energy in mode 7's eigenfunction is captured by
the vertical aperture of the LVLA. Mode 7 is the highest order mode at 5 Hz whose
lower turning point is still within the water column.

A generalized matrix inverse approach that provides a least-squares best fit for
the mode amplitudes and that yields generalized mode eigenfunctions can be derived
(TINDLE *et al.*, 1978). The approach was applied to the LVLA recordings of the
VAST *T*-phase events. It does provide a correction for the discrete sampling in depth
performed by the array (TINDLE *et al.*, 1978). However, when the array does not span
the full waveguide, the required matrix inversion becomes increasingly less stable as

Orthogonality of Modes across LVLA Array

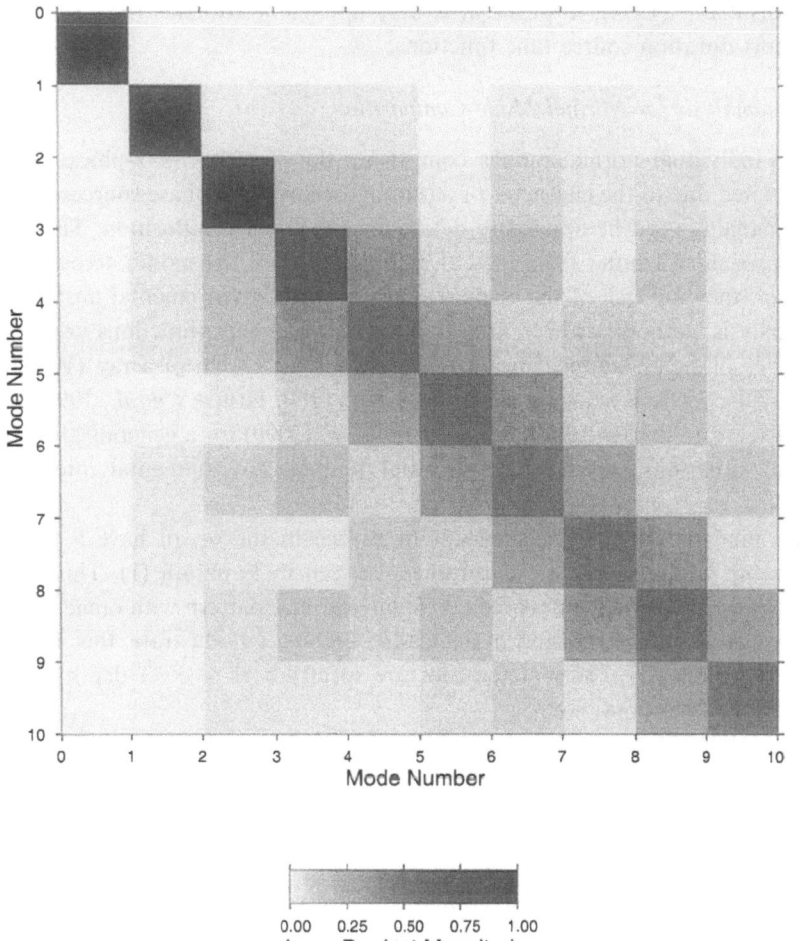

Figure 10

Value of the inner product of the first 10 orthonormalized mode eigenfunctions at 5 Hz, evaluated at the 48 depths corresponding to those of the LVLA elements, with each of the other 10 modes. The mode eigenfunctions were calculated by the Kraken computer code.

the number of modes considered increases. At a frequency of 5 Hz, when just the first four or five modes were considered, the matrix inversion yielded generalized mode eigenfunctions that were nearly identical to the original eigenfunctions and eigenvalue levels consistent with those in Figure 9. As the number of modes were increased, the generalized eigenfunctions deviated by an increasingly greater extent from the original ones. Because of this dependence on the number of modes considered, the results of this method are not presented here. A method of mode filtering with data from arrays with small vertical aperture is presented in HEANEY and KUPERMAN (1998); however, it is only applicable to fields from point sources with short-duration source time functions.

C. Eigenanalysis for Normal Mode Composition

The individual normal modes comprising the waterborne T-phase energy are uncorrelated due to the character of naturally-occurring T-phase sources. Therefore, an eigenanalysis can be applied to determine the mode composition. The results of this approach still suffer from lack of orthogonality of the modes across the array. However, they are free of the errors associated with environmental mismatch. This eigenanalysis method has been used to estimate mode eigenfunctions using only the properties of ocean ambient noise recorded across a vertical array (WOLF, 1987; WOLF *et al.*, 1993; KUPERMAN and INGENITO, 1980; HURSKY *et al.*, 1995; NEILSEN *et al.*, 1997; and SMITH, 1997). See HURSKY *et al.* (2000) for a generalization of these ideas to situations where additional, but limited, environmental information is available.

The modes excited by a single point source in the ocean have a fixed phase relationship with respect to one another; as seen in Equation (1). This intermode phase relationship results in a predictable interference pattern with range, depth, and frequency (e.g., D'SPAIN and KUPERMAN, 1999). To illustrate this interference pattern more clearly, consider the pressure spectrum at receiver depth z_j due to a single underwater point source:

$$\langle p(z_j)p^*(z_j)\rangle = P(\omega)\sum_n a_n a_n^* [\Psi_n^s(z_s)\Psi_n^r(z_j)]^2$$

$$+ 2P(\omega)\sum_{n,m}[a_n a_n^* a_m a_m^*]^{1/2}\Psi_n^s(z_s)\Psi_m^s(z_s)\Psi_n^r(z)\Psi_m^r(z)\cos(\Delta\bar{\kappa}_{nm}(\omega)r) \quad (5)$$

where $P(\omega) \equiv A(\omega)A^*(\omega)$ is the source pressure spectrum and $\Delta\bar{\kappa}_{nm}$ is the range-averaged difference in mode wavenumbers. The received spectrum is composed of two terms; the first is only weakly dependent upon range, whereas the second term is oscillatory in range. The first term is the incoherent summation of the mode energy and the second represents the interference pattern between modes.

When the modes are strongly correlated with one another, an eigenanalysis of the array data cross spectral matrix yields information about the sources contributing to the field; the number of significant eigenvalues equals the number of point sources, and their corresponding eigenvectors are generalized direction vectors to the source (SCHMIDT, 1981). In addition, the intermode interference pattern contains useful information such as the range to the source. However, for T-phase arrivals from naturally occurring events, the situation is different. Due to the nature of the seismic source and the coupling of its energy into the water column, the contributions to the received ocean acoustic field by any mode at any time during the T-phase arrival come from several different ranges. Integrating these various contributions over the source region results in the incoherent enhancement of the first term in Equation (5) and a cancellation of the second term. Given that the range extent of the source region is on the order of an intermode interference wavelength, $2\pi/\Delta\bar{\kappa}_{mn}$, or greater, then the second term will be negligible with respect to the first term. In addition, the intermode phase relationship at a given frequency is lost to a great extent during propagation in the solid earth along the different paths that the mode energy must take to couple into the water column at the appropriate water depths. Also, both the averaging over time that is done to obtain a stable estimate of the data cross spectral matrix, and the long ranges to the sources from the receiving array help to decorrelate the modes (WOLF et al., 1993).

The cross spectrum between the pressure fields received by two vertical array elements at depths z_j and z_k also has terms containing $\Delta\kappa_{mn}r$ which tend to zero as the contributions from various ranges are summed. For this reason, and for the additional reasons listed in the previous paragraph, the elements of the pressure cross spectral matrix are of the form:

$$\langle P_{tot}(z_j)p_{tot}^*(z_k)\rangle \approx \sum_n B_n^2(\omega)\Psi_n^r(z_j)\Psi_n^r(z_k) \qquad (6)$$

where B_n is the range-integrated magnitude of mode n. Therefore, the pressure data cross spectral matrix across the array, $[\mathbf{S}(\omega)]$, can be written as:

$$[\mathbf{S}(\omega)] = \sum_n B_n^2(\omega)\underline{\Psi}_n\underline{\Psi}_n^T = [\mathbf{E}]_{M\times N}[\mathbf{B}(\omega)]_{N\times N}[\mathbf{E}]_{N\times M}^T \qquad (7)$$

where $[\mathbf{E}]_{M\times N}$ is the matrix that has as its columns the $(M \times 1)$ mode eigenfunction vectors $\underline{\Psi}_n$ defined earlier, and $[\mathbf{B}(\omega)]$ is a diagonal matrix with the mode squared amplitudes, $B_j^2, j = 1, \ldots, n$, along the diagonal. Equation (7) has exactly the form of an eigenvector/eigenvalue decomposition of a Hermitian matrix (NOBLE, 1969). Therefore, given that the modes are mutually uncorrelated, the eigenvectors of the pressure data cross spectral matrix of a vertical array are associated with the mode eigenfunctions and the eigenvalues with the modes' mean-square amplitudes.

The pressure data cross spectral matrix is Hermitian so that eigenvectors corresponding to distinct eigenvalues are necessarily orthogonal (NOBLE, 1969). However, these eigenvectors do not necessarily correspond exactly to the mode eigenfunctions; they do so only to the extent that the modes are orthogonal across the array aperture. Lack of mode orthogonality causes the eigenvalues to represent the energy in more than a single normal mode and the corresponding eigenfunctions to be distorted hybrids of the true mode eigenfunctions. Other sources of error, e.g., variations in sensitivity of the array elements, map into distortions in the resulting mode eigenfunctions. However, with this technique, no assumptions about, nor prior knowledge of, the environmental conditions need to be made, only that the arriving normal modes are mutually uncorrelated.

Figure 11 is a plot of the normalized level of the ten largest eigenvalues for the integral frequencies from 3 to 10 Hz for the offshore event at Mendocino Ridge.

It shows that the four largest eigenvalues are those at 3 to 6 Hz, an indication of the predominant frequency content of the arrivals. Another aspect of the plot is the regular decrease in eigenvalue level with frequency. At 3 Hz, the three largest eigenvalues show a given rate of decrease in level with increasing eigenvalue number, followed by a rapid dropoff. At 4 Hz, the dropoff occurs after the 4th eigenvalue, at 5 Hz, after the 4th and 5th eigenvalues, etc. Numerical simulations indicate that a rapid dropoff in eigenvalue levels occurs because of the lack of mode orthogonality across the vertical aperture of the LVLA. However, the dropoff typically occurs at higher eigenvalue numbers than shown in this plot. Using the simulations as a guide,

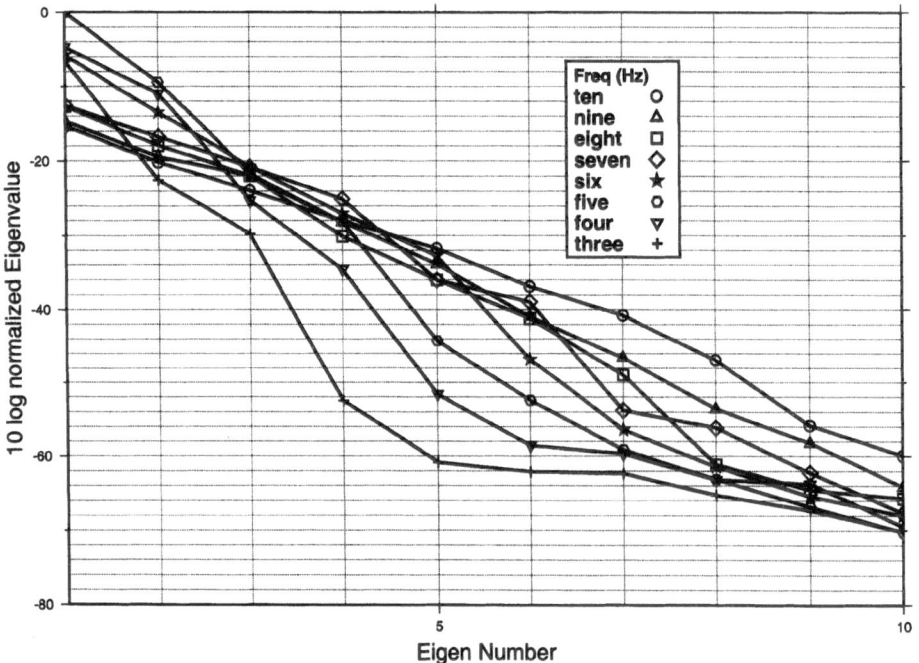

Figure 11

Plot of the normalized level of the ten largest eigenvalues for each of the eight data cross spectral matrices at integral frequencies from 3 to 10 Hz for the offshore event 3 at Mendocino Ridge. The cross spectral matrices were calculated by dividing 25.6 s of time series data encompassing the highest-level signals of the T-phase arrival into 19 segments, each overlapped by 50 percent, Fourier transforming each segment after windowing with a Kaiser-Bessel window of α equal to 2.5, calculating either the autospectrum or cross spectrum, and then incoherently averaging the resulting 19 snapshots. This procedure provides more than 37 degrees of freedom for the estimation of the matrix elements and a frequency resolution of 0.39 Hz. An eigenanalysis then was performed on the matrices at each integral frequency. The largest eigenvalue across frequency, the one at 5 Hz, was used to normalize the eigenvalues at all frequencies, both in this figure and in Figure 13. The results then were converted to decibel units by taking ten times the logarithm.

the number of eigenvalues before and during the rapid dropoff at a given frequency can be used as an indication of the number of significant normal modes constituting the arrivals at this frequency. In particular, the number of significant modes in the *T*-phase arrival at 5 Hz is four to six. This result is consistent with the conventional mode decomposition results shown in Figure 9, particularly after accounting for the rolloff in levels along the diagonal in Figure 10.

The corresponding eigenvectors for the five largest eigenvalues at 5 Hz for the offshore event are shown in Figure 12. These eigenvectors, are similar in appearance to the normal mode eigenfunction amplitudes. At this frequency and over this vertical depth span, the normal mode number is equal to the number of humps in the magnitude versus depth plot. Therefore, the eigenvector corresponding to the largest eigenvalue (connected circles in Fig. 12) has four humps and so must represent mode 4. The eigenvector for the second largest eigenvalue represents mode 3, the third largest represents mode 2, the fourth mode 5, and the last one represents mode 1. This ordering of the relative mode contributions to event 3's arrival again is consistent with that obtained from Figure 9. Note that the excessively large

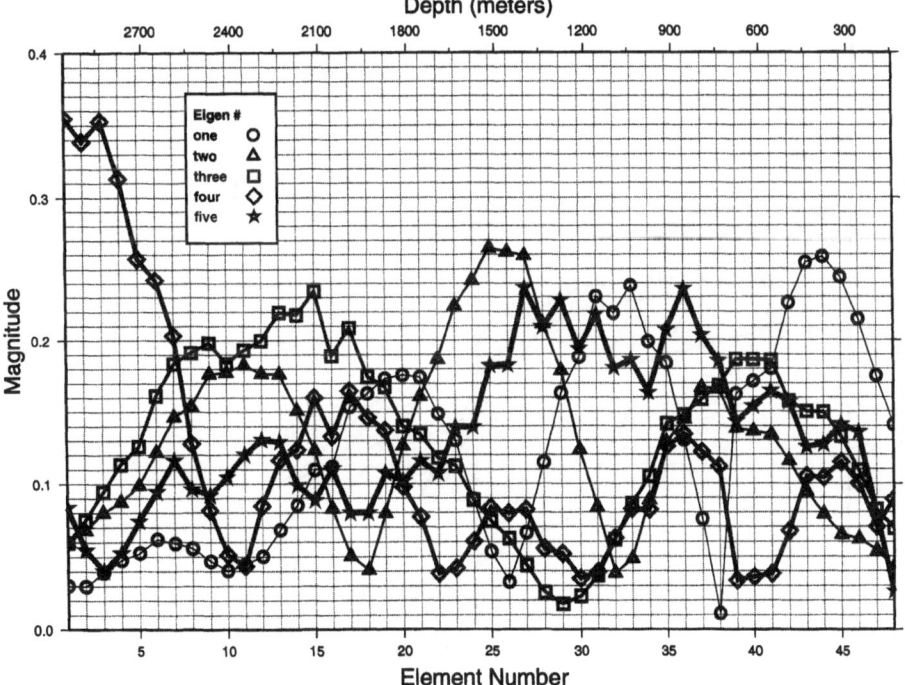

Figure 12

Plot of the five eigenvectors corresponding to the five largest eigenvalues at 5 Hz as a function of depth (element number) across the LVLA for the *T*-phase arrival from the offshore event 3. The procedure used for the eigenanalysis is described in the caption for Figure 11.

magnitude of the eigenvector representing mode 5 near the bottom of the array in Figure 12 (connected diamonds) may be indicative of some degree of leakage between modes 4 and 5, and may explain the low level of the hump at greatest depth in the eigenvector representing mode 4.

As previously mentioned, these eigenvectors are only approximations to the actual mode eigenfunctions, due mostly to lack of mode orthogonality across the array aperture.

The corresponding plot of the largest eigenvalues for the integral frequencies from 3 to 10 Hz for the land-based event near San Jose, California is given in Figure 13. The regular increase in the number of significant eigenvalues with increasing frequency again is apparent. However, the point of the sudden decrease in level occurs at slightly smaller eigenvalue numbers at most frequencies than seen in event 3, suggesting that fewer modes contribute to this arrival. Also, the largest eigenvalues are about 20 dB lower than the largest one in event 3 and show a smaller spread in level from 3 to 7 Hz.

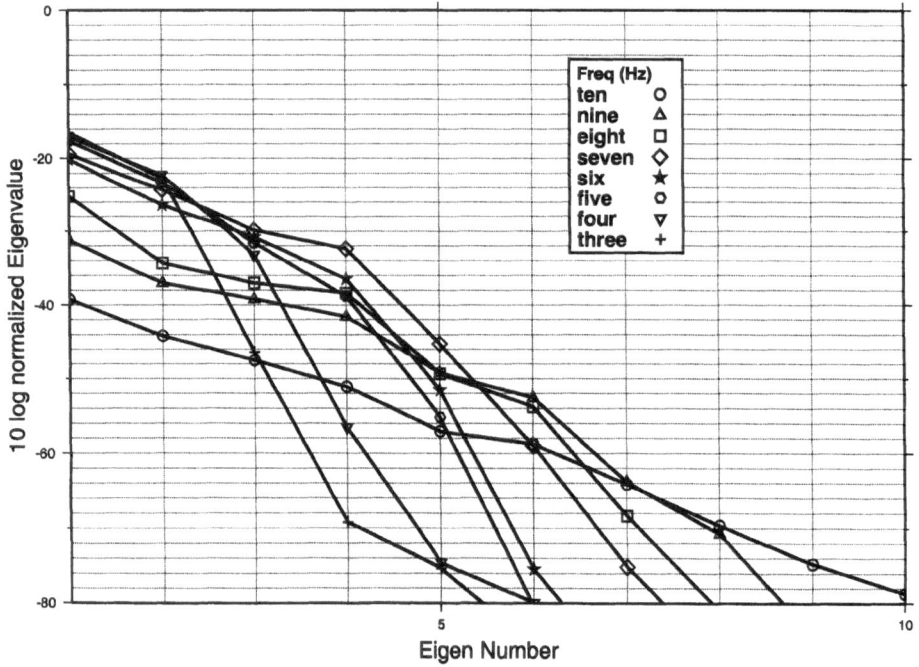

Figure 13
Plot of the normalized level of the ten largest eigenvalues for each of the eight data cross spectral matrices at integral frequencies from 3 to 10 Hz for event 4 near San Jose, California. The normalization factor for all frequencies is the largest eigenvalue for the offshore event in Figure 11, i.e., the one at 5 Hz. The method of calculating and plotting these results is the same as in Figure 11.

The eigenvectors associated with the five largest eigenvalues at 5 Hz for this event are plotted in Figure 14. The one for the largest eigenvalue clearly represents a significantly lower order mode than that in the offshore event. The major hump for this eigenvector is broadly centered around the sound channel axis and is strongly suggestive of the lowest order mode 1's eigenfunction, although a secondary hump does occur around 2000-m depth. Note that the eigenvectors for the three largest eigenvalues all display two humps across the array aperture. Because mode 2 has a lower turning point between 1800 and 2000 m, the eigenvector corresponding to the third largest eigenvalue (connected squares) best represents mode 2, with the second largest eigenvalue's eigenvector corresponding to mode 3. The fourth and fifth largest eigenvalues' eigenvectors represent modes 4 and 5, respectively.

In any case, the dominant modes at 5 Hz in event 4's arrival are certainly of lower order than those in event 3, in agreement with the conventional mode decomposition results in Figure 9. This difference in mode content is consistent with the coupling model presented in the Introduction; for land-based events, the seismic energy that

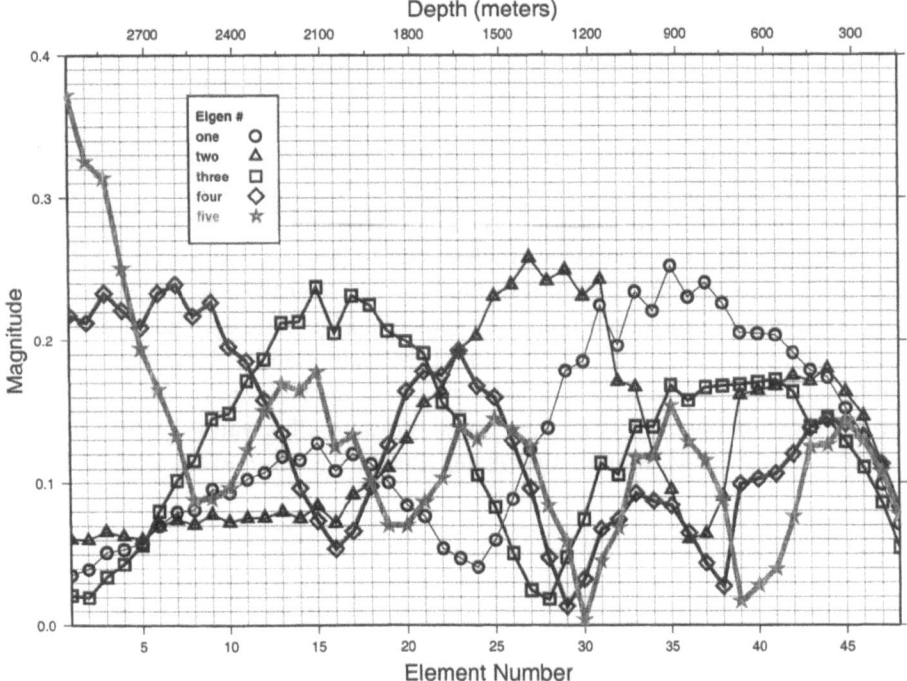

Figure 14

Plot of the five eigenvectors corresponding to the five largest eigenvalues at 5 Hz as a function of depth (element number) across the LVLA for the *T*-phase arrival from event 4 on land. Other aspects of this plot are the same as in Figure 12.

couples into the higher order modes must propagate for a longer distance through the lossy earth before coupling occurs.

II. Mode Composition from Single Hydrophone Data

Receiving systems with vertical aperture are not part of the International Monitoring System and so the processing described in the previous section is not applicable to IMS data. However, under certain circumstances, it may be possible to obtain a crude estimate of the mode content of arrivals using data from a single underwater hydrophone. These circumstances occur when a short-duration, in-water point source with sufficient low frequency content, e.g., an in-water detonation, is recorded at long range by an underwater receiver and the propagation path is solely within the ocean. The purpose of this section is to present the ideas behind this possibility. In the process, the arrival structure of some interesting events such as naturally-occurring events at Hawaii and the French nuclear tests at Mururoa and Fagataufa is shown to be partially explained by the dispersion characteristics of deep water propagation. Once these signals couple into land vibrations, however, their arrival structure is strongly altered by the coupling characteristics.

A. T Phase Recordings of Hawaiian Events at Pt. Sur

Recordings of the *T*-phase arrivals from the four largest Hawaiian events in 1996 and 1997 have been acquired from the offshore Pt. Sur station and several land stations in western North America. Information pertaining to these four events, extracted from the Reviewed Events Bulletin (REB), is contained in Table 2. The last column also provides the epicentral distance of the event from the Pt. Sur receiver.

The locations of these four events are shown in Figure 15. The first two occurred at the Loihi Seamount on the south flank of the main island of Hawaii and events 3 and 4 took place on Hawaii itself. Also shown on the map are the locations of the two 1996 events obtained from the Hawaii Volcano Observatory (HVO) network data (CAPLAN-AUERBACH and DUENNEBIER, 2000). The offset of the REB and HVO

Table 2

Four Hawaiian earthquakes analyzed in Pt. Sur Data, 1996 and 1997

Quake no.	REB ORID	Year	JD	Time	Latit.	Longit.	m_b	Range (km)
1	751298	1996	205	13:24:59	18.9620	−155.3959	4.40	3753
2	752002	1996	206	17:38:50	18.9833	−155.4213	4.51	3754
3	1074833	1997	181	15:47:38	19.3108	−155.1058	5.06	3706
4	1110057	1997	227	01:54:38	19.4083	−155.1098	4.43	3699

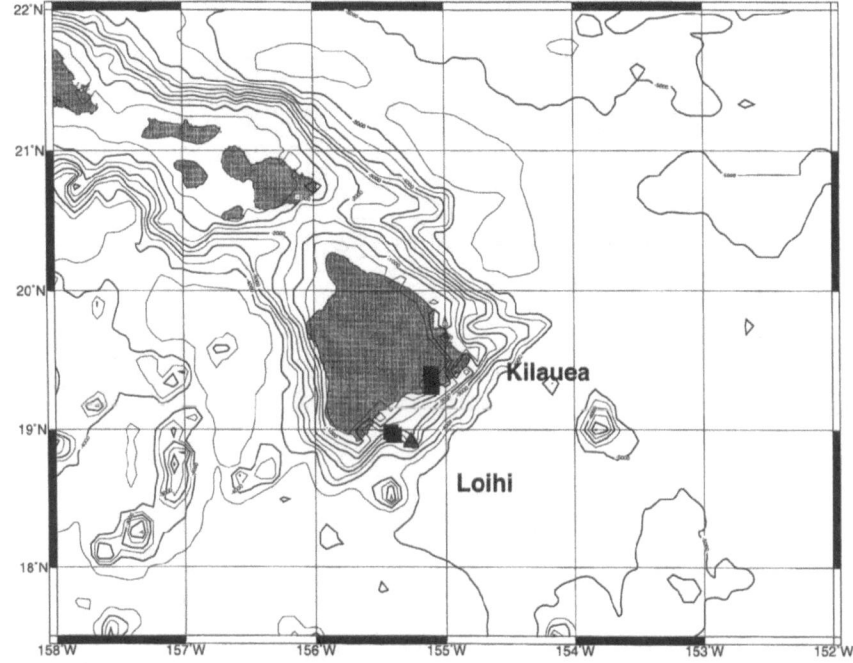

Figure 15

Map of the Hawaiian islands and surrounding ocean bottom bathymetry. The squares labeled 1 through 4 (the square for event 2 is covered by that for event 1) show the epicentral locations of Hawaiian events 1 through 4 given in Table 2 as obtained from the Reviewed Events Bulletin (PROTOTYPE INTERNATIONAL DATA CENTER, 1998). The two triangles (which are co-located on this spatial scale) indicate the corresponding epicentral locations for events 1 and 2 in 1996 derived from Hawaii Volcano Observatory local station recordings (CAPLAN-AUERBACH and DUENNEBIER, 2000). The bathymetry information, obtained from the GMT data base (NATIONAL GEOPHYSICAL DATA CENTER, 1998), is contoured in 500-m increments.

positions is less than 20 km. In addition, the estimated origin times of the two events in the REB and the HVO-derived results differ by less than 2 s.

Events 1 and 4 in Table 2 have characteristics much in common with the pair of VAST earthquakes examined in the previous section. Both events have nearly identical body wave magnitudes and occur in the same direction, and at the same epicentral distance, from the Pt. Sur station. Whereas event 1 occurred offshore below an approximately 1500-m-deep water column, event 4 occurred on land. Figures 16 and 17 are plots of the contoured spectral ratio spectrograms for these two events as recorded by the Pt. Sur hydrophone. The spectral ratio spectrogram for event 2 is very similar to that of event 1 in Figure 16, and the gram for event 3 is very similar to that of event 4 in Figure 17 except that its signal-to-noise ratio is much higher. Both Figures 16 and 17 show a main arrival followed a minute or so later by a secondary arrival with a time-frequency structure very

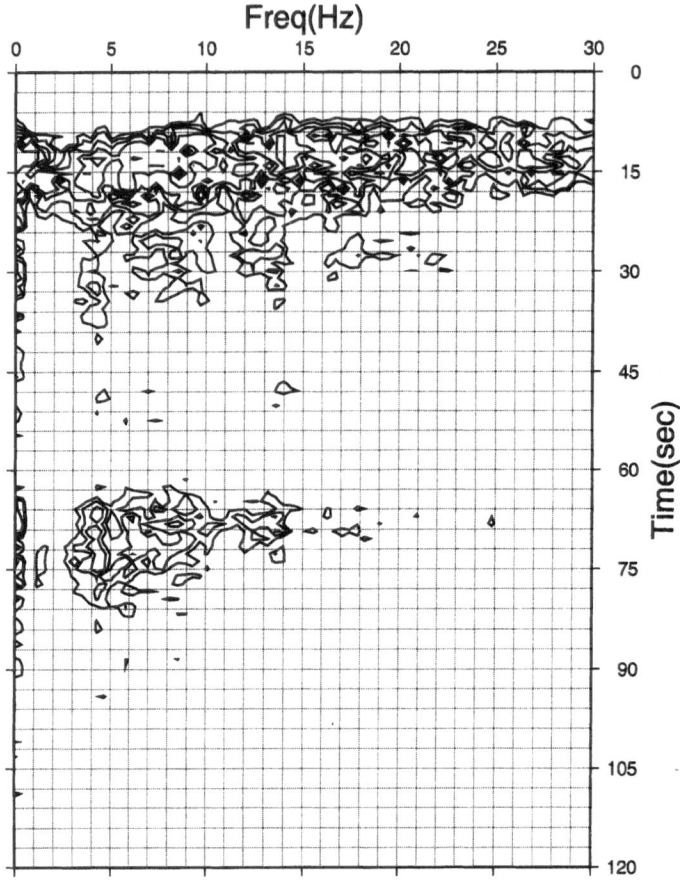

Figure 16

Contour plot of the time-frequency spectral ratio spectrogram from 2 min. of Pt. Sur hydrophone data over the band from 0 to 30 Hz for the m_b 4.40 Loihi event 1 listed in Table 2 of the text. The spectral ratio was calculated by estimating the noise spectral density from 10 s of data prior to the main T-phase arrival (providing seven statistically independent estimates for the incoherent average), and using it to normalize the spectral densities estimated during the period shown in the plot. This procedure eliminates the need to account for the data acquisition system response. The Fourier transforms have a frequency resolution of 0.4 Hz and a temporal resolution of 1.3 s. The contours occur in 6 dB steps from 22 dB to 46 dB.

much like that of the main arrival. Analysis not shown here indicates that the spectrum of the secondary arrival decreases somewhat more rapidly with increasing frequency than the main arrival. In addition, the higher frequency portion of both arrivals has a remarkably shorter duration (about 15 s) than the 1–2 min. duration of the T-phase events recorded in the VAST experiment. About half of this duration (5–7 s) is associated with dispersive propagation over the deep water path (Section IIB). Also, both figures indicate a spectral trough centered about 1 or 2 Hz.

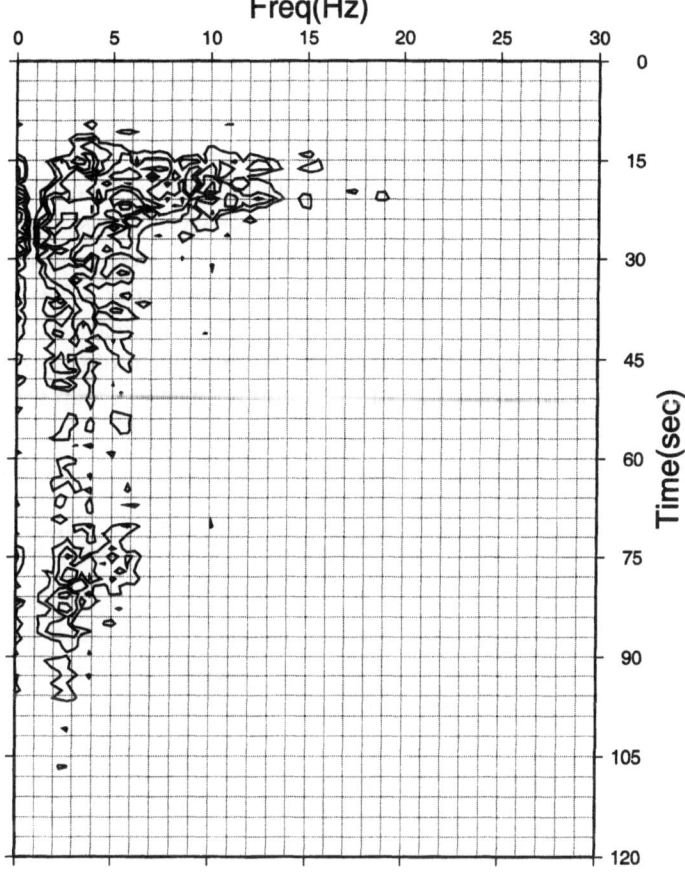

Figure 17

Contour plot of the spectral-ratio spectrogram estimated from Pt. Sur hydrophone data for the m_b 4.43 Hawaiian event 4 listed in Table 2. All other details of the plot including the levels of the contours are analogous to those in Figure 16.

The arrival structure for the two events differs in interesting ways. First, the offshore event has higher frequency content than the one on land, just as observed with the pair of VAST events. Second, the leading edge of event 1 is composed of higher frequencies, with frequencies below about 8 Hz arriving a few seconds later. In contrast, event 4's leading edge contains 3–4 Hz energy with the higher frequencies coming in up to 4 s later. A third difference is that the coda for the on-land event is composed of significantly lower frequencies (1–6 Hz) than that for the offshore event (4–20 Hz). Finally, the duration of the coda for the offshore event is only about 75 percent of the duration of the low frequency tail in event 4, even after taking into account the slight difference in body-wave magnitude.

B. T-wave Dispersion over the Oceanic Path

The mode group velocities as a function of frequency were calculated for the great circle path from Loihi to Pt. Sur. These calculations were performed with the Kraken normal mode code (PORTER, 1991) in the following way. First, the 3746-km path from Loihi to Pt. Sur was divided into 38 range-independent sectors, based on the bottom bathymetry (obtained from NATIONAL GEOPHYSICAL DATA CENTER, 1998) and the sound speed profile variations (from NEMO OCEANOGRAPHIC DATA SERVER, 1998) along the path. The mode group velocities were calculated for each sector using perturbation theory (Eq. (5.150) of JENSEN *et al.*, 1994). Using the adiabatic approximation, the mode group slownesses for each sector then were averaged, after weighting by the percentage of the range extent of the sector over the full path distance. Figure 18 is a plot of the resulting effective group velocity versus frequency for the first 10 modes for the first 3710 km of the path from Loihi to the bottom of the continental slope leading to Pt. Sur. The results in this figure predict

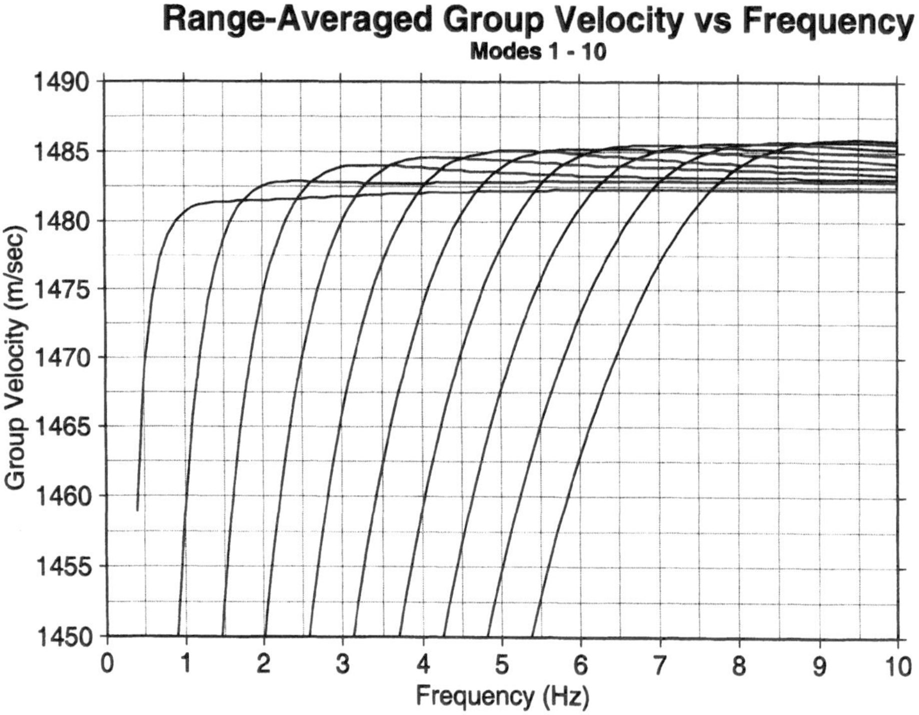

Range-Averaged Group Velocity vs Frequency
Modes 1 - 10

Figure 18
Plot of the range-averaged mode group velocities as a function of frequency for the first 10 normal modes for the first 3710 km of the great circle path (the first 18 range-independent sectors) from Loihi to the start of the continental slope leading to Pt. Sur. For a given epicentral distance, the figure's vertical axis can be converted into travel time; the time of arrival difference between energy traveling over 3710 km at 1485 m/s and at 1450 m/s is 60 s. Details of the calculations are presented in the text.

that a single-impulse source at 3710-km range that excites the first 10 normal modes will give rise to a broadband first arrival with nearly a 7 s duration (due to the 1482–1486 m/s spread in group velocities) followed by a long duration coda with lower frequency content. This predicted arrival structure matches quite well the general character of the single element spectrograms in Figures 16 and 17. In fact, the comparison indicates that a rough estimate of the number of significant modes in the Hawaiian arrivals is obtainable from these single element data by measuring the bandwidth of the arrivals several seconds, say 20 s, after the initial arrival. If this interpretation of the Pt. Sur data is correct, it indicates that the energy in the offshore event in Figure 16 is composed of a broader set of modes than the on-land event, but is somewhat deficient in the lowest order modes. In contrast, the on-land event's arrival structure, in Figure 17, is dominated by the lowest few modes, particularly its long-duration, low frequency coda. This interpretation of the differences in arrival structure between the offshore and the on-land events coincides with the differences between the mode content of the two VAST earthquake arrivals discussed previously.

The span of 1482–1486 m/s in the group velocities of the broadband initial arrival in Figure 18 is associated with the characteristics of deep sound channel propagation. Since the acoustic waveguide in this case is formed by refraction, the group velocity typically increases with increasing mode number at a fixed frequency and with decreasing frequency for a given mode. The reason for this behavior is that the mode energy is able to penetrate to greater distances from the sound channel axis where the medium sound speeds are higher. This increase in medium sound speed is sufficiently great in most deep water environments to more than compensate for the increase in equivalent ray path distance. For example, the fourth mode in Figure 18 has its greatest group velocity at 4 Hz where it has the greatest vertical extent in the water column without significant interaction with the bottom. For increasing frequencies above this point, the group velocity decreases slightly since the mode's vertical distribution of energy becomes progressively more concentrated about the sound channel axis depth. At decreasing frequencies below 4 Hz, the mode interacts an increasing amount with the ocean bottom. This interaction causes a remarkable slowing of the mode's group speed. In effect, the mode is dragging its eigenfunction tail along the ocean bottom, giving rise to a predicted low frequency coda, following the broadband initial arrival. At these lowest frequencies, mode propagation is equivalent to propagation in shallow water environments where the acoustic waveguide is formed by reflection from both upper and lower boundaries, rather than refraction.

Figure 19 presents the effective group velocity results in which all 38 sectors over the complete 3746-km distance from Loihi to Pt. Sur are included. Two significant differences exist between this figure and the previous one. First, the span of group velocity values in the broadband leading arrival decreases from 1482–1486 m/s to 1482–1485 m/s. This decrease is indicative of a "recompression" of a deep-water-dispersed arrival due to bottom interaction along the continental slope (BURENKOV,

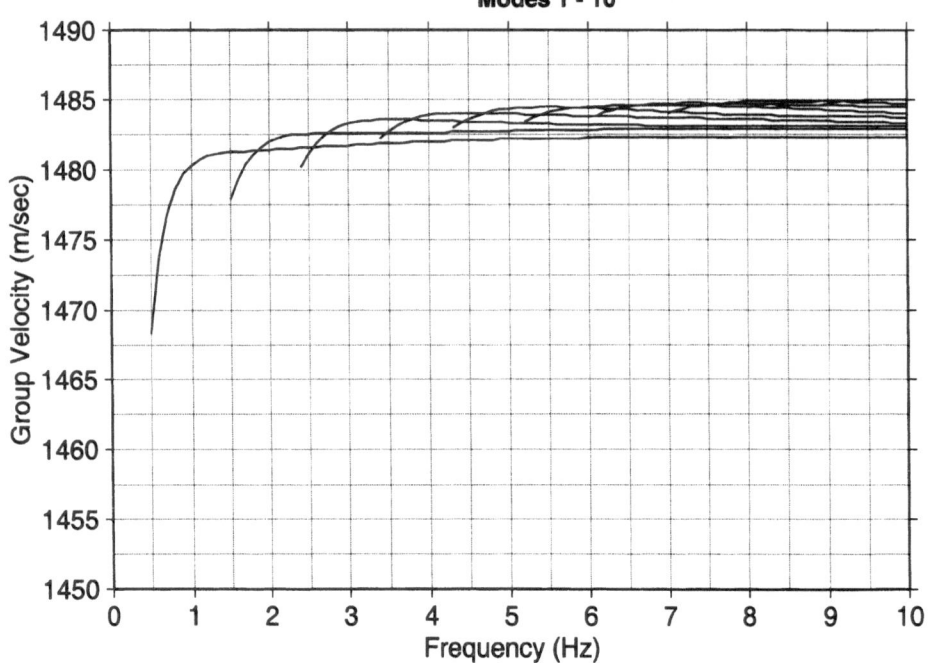

Figure 19
Plot of the range-averaged mode group velocities as a function of frequency for the first 10 normal modes
over the complete 3746 km great circle path from Loihi to Pt. Sur.

1989). That is, propagation into the increasingly shallower water portion of the path, where the mode group velocities now *decrease* with increasing mode number and decreasing frequency, partially undoes the dispersive effects in the deep water portion. The importance of shallow water dispersion along the sloping part of the propagation path is discussed in JOHNSON (1963).

The second difference between Figure 19 and Figure 18 is that propagation at the smallest values of the modes' group velocities is no longer allowed by the shoaling bottom bathymetry. This result suggests that the low frequency coda is extinguished to a large extent by the continental slope before reaching the Pt. Sur hydrophone. However, the process of mode cutoff is more interesting and more involved than the sudden disappearance of modes illustrated in Figure 19 (e.g., COPPENS and SANDERS, 1978; JENSEN and KUPERMAN, 1980). A given mode continues to contribute to the sounds recorded by a water column receiver, particularly one on the ocean bottom, not only as it reaches its cutoff depth, but also for a significant range thereafter. The properties of the received signal on land are strongly dictated by the ocean bottom geoacoustic properties over the span of ranges where mode cutoff and coupling occurs.

The effects of deep water dispersion also appear in the arrival structure of the 1995 French Polynesia nuclear tests recorded at Pt. Sur (PROTOTYPE INTERNATIONAL DATA CENTER, 1998). Figure 20 shows a plot of the spectral ratio spectrogram for the 27 October event at the Mururoa Atoll. This event's time-frequency arrival structure is quite similar to that of the on-land Hawaiian events (Fig. 17), except that its secondary arrival trails the main arrival by only about 50 s and it does not have an appearance similar to the main arrival. The 50-s arrival time difference is consistent with scattering/reflection off the Marquesas Islands. Although the land-to-water coupling at the source site must certainly have a distorting effect on the received waveforms, dispersion in the deep ocean sound channel over the 6670 km path also

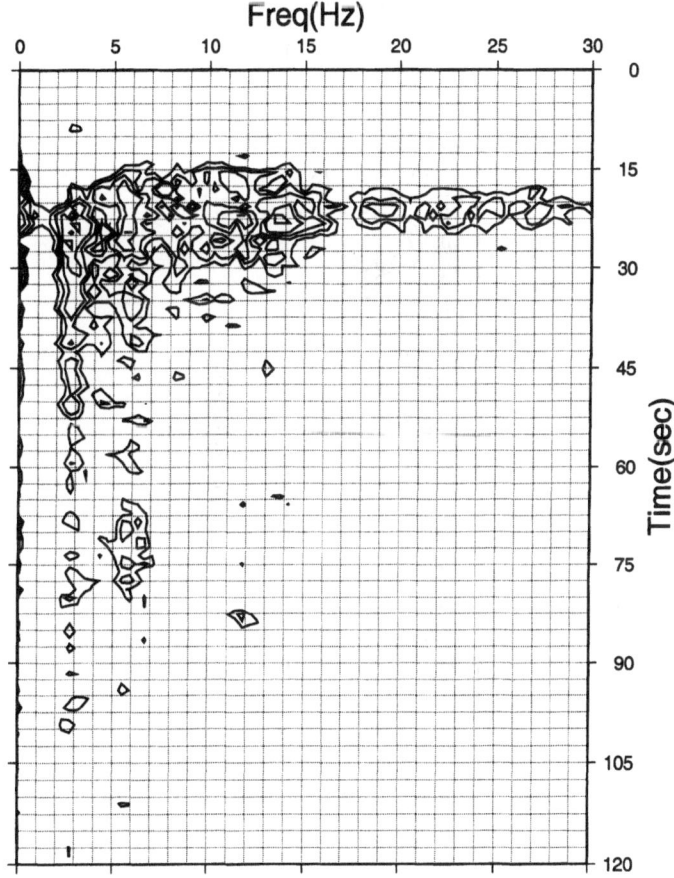

Figure 20
Contour plot of the spectral ratio spectrogram for the 27, October, 1995, French nuclear test on the Mururoa Atoll, as recorded by the Pt. Sur hydrophone. This event had an announced yield of 60 ktons (PROTOTYPE INTERNATIONAL DATA CENTER, 1998). Other features of this plot are analogous to those in Figures 16 and 17.

clearly plays a role. The results in Figure 20 suggest that *T*-phase arrival times are best picked at the highest frequencies where the waveforms are least spread out in time. The use of the high frequency portion of the *T*-phase signal for picking arrival times, as well as the effects of dispersion on *T*-phase arrival structure, are discussed in NORTHROP (1962).

C. *Land-based Recordings of Hawaiian Events*

The arrival structure of the Hawaiian events recorded on land exhibits distinctively different characteristics than the in-water recordings. Figure 21 presents

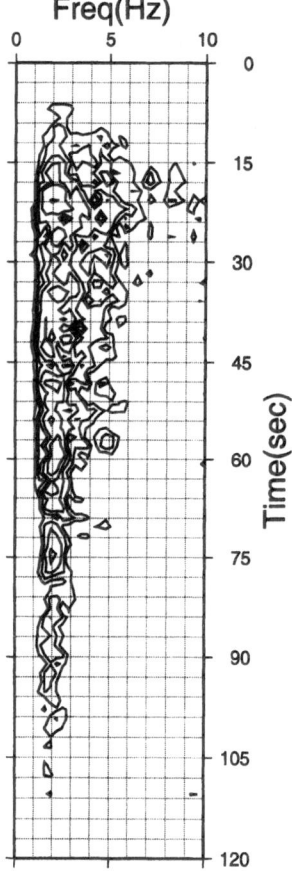

Figure 21
Time-frequency spectral ratio contour plot for event 3, at Loihi in 1997, as recorded by the vertical component at station BKS (Byerly Vault, Berkeley, California). For this plot, the noise spectral density was estimated from nearly 13 s of data prior to the main arrival. The Fourier transforms have a frequency resolution of 0.3 Hz and a temporal resolution of 1.6 s. The contours are plotted in 6 dB steps from 10 dB to 34 dB.

the spectral ratio spectrogram for the vertical component recorded at the Berkeley station BKS for the magnitude 5.06 event 3, and the spectral ratio spectrogram for the magnitude 4.43 event 4 recorded by the *T*-phase station VIB is given in Figure 22. The contour levels are 10 dB to 34 dB in both figures. The BKS recording of event 3 is presented rather than that of event 4 because the higher *T*-phase levels from the larger magnitude event are required to be clearly visible on this contour level scale. The corresponding contour plot of the m_b 5.06 event at VIB has an arrival structure nearly identical to that of event 4 in Figure 22 once the contour levels are increased by about 30 dB.

A significant difference between the arrival structure in the water column and on land is that much of the higher frequency content of the broadband arrival is filtered out, either by water-land coupling or by propagation on land, or both. Although the

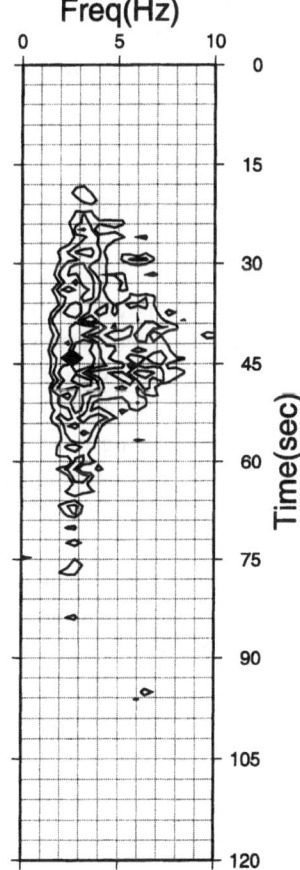

Figure 22

Spectral ratio spectrogram contour plot from VIB (Van Inlet, Canada) for event 4, the body-wave magnitude 4.43 Hawaiian event in 1997. The contours are plotted in 6 dB steps from 10 dB to 34 dB.

VIB data were digitized at 100 samples, Figure 22 extends only up to 10 Hz because no additional energy on this contour scale appears at higher frequencies. (The BKS data were sampled at 20 samples.) This low-pass filtering effect also is clearly evident in the comparison of the two VAST events (Figs. 3 and 4) and the two Hawaiian events (Figs. 16 and 17). The remnants of a broadband arrival can be identified at BKS about 10 s after the initially arriving energy in the 2–3 Hz band. This delay from the initial onset is about twice that seen in the Pt. Sur recordings for this Loihi event (nearly identical to Fig. 17). The delay to the broadband arrival at VIB is even greater, about 2.5 times that at BKS. As seen in Figure 22, the arrival structure displays a taper, starting with initial energy in the 3–5 Hz band, to a broader band of arrivals about 25 s later. These differences in the land and water column arrival structure are associated with the fact that the lower frequencies couple into the land seismic field at farther ranges offshore, and so travel a greater distance at the higher seismic velocities. (See also TOLSTOY and EWING, 1950.) The water-to-land coupling confuses the relative time-of-arrival of the modes, resulting in loss of information. Numerical modeling of these effects, and quantitative comparison with the data presented here, is the focus of future work.

III. Conclusions

Naturally-occurring T phases can be used as probes of opportunity to study the characteristics of the coupling of the underwater acoustic field into the land seismic field. Their use for this purpose is greatly aided by an understanding of the properties of the T phases themselves. In particular, their normal mode composition determines to a great extent how and where along the continental shelf and slope they couple into land vibrations. This insight into the coupling process is provided by the classical model of upslope sound propagation in a wedge-shaped ocean. It describes, at least qualitatively, many of the features seen in T-phase recordings. The reciprocal view of the model provides one mechanism for the coupling of land-based vibrations into the underwater acoustic field and provides an explanation for the time-dependent vertical arrival structure seen in the T-phase recordings from events on land.

Data recorded in VAST experiment by a 3000-m vertical hydrophone array are used to determine the normal mode components of T-phase recordings from five earthquakes. Results are presented and discussed for two events at about the same range and in the same direction from the array and which have equivalent body-wave magnitudes of 4.1. One event occurred offshore where the overlying water was about 3000-m deep, whereas the second event occurred on land. Comparison of the mode decomposition results at 5 Hz indicates that the relative energy in each of the modes is dependent upon coupling characteristics at the source, with the offshore event having higher order mode content than the event on land. The normal modes propagate in an uncorrelated way with one another due to the significant range extent of the source region and to the manner in which the source seismic energy couples into the water column. This uncorrelated nature of the mode propagation implies that efforts at numerical modeling of the T-phase arrival structure from naturally-occurring events need only predict amplitude envelopes (as in DE GROOT-HEDLIN, 1998; DE GROOT-HEDLIN and ORCUTT, 1999) rather than the intermode interference phenomena observed in underwater recordings of in-water point sources. Likewise, only the magnitude of land-to-water transfer functions is relevant. The uncorrelated nature of the modes permits a simple eigenanalysis of the array data cross spectral matrix to reveal the normal mode content so that the mode eigenfunctions need not be calculated numerically, as is required with conventional mode decomposition. The eigenanalysis and conventional mode decomposition results of the VAST events are consistent with one another.

The lack of correlation between the normal modes in naturally-occurring T-phase arrivals has implications for use of T-phase stations in the IMS. That is, mode interference patterns (or relative time of arrival between the various modes) recorded underwater from an underwater point source contain important information such as the source range (e.g., D'SPAIN and KUPERMAN, 1999). However, the characteristics of water-to-land coupling confuse the relative phase relationship (timing) between modes. This loss of information occurs along with the decrease in signal amplitude, particularly at higher frequencies, associated with the coupling.

The time-frequency arrival structure at the Pt. Sur hydrophone station of the 1995 nuclear tests in French Polynesia and the Hawaiian/Loihi events in 1996 and 1997 can be explained to a large extent by the range-averaged dispersion characteristics along the ocean path. Alternative explanations associated with propagation along the initial land path have been presented (TALANDIER and OKAL, 1998). Given that the dispersive character of ocean path propagation is the dominant effect in determining the arrival structure, it provides a way of crudely estimating the number of significant normal modes, using data from a single, omnidirectional, in-water receiver. This approach is most likely to be valid when a short-duration, in-water point source with sufficient low frequency content, e.g., an in-water detonation, is recorded at long range by an underwater receiver and the propagation path is solely within the ocean. Differences in arrival structure for two m_b 4.4 Hawaiian events, one occurring offshore and the other on land, suggest differences in normal mode content consistent with those seen in the pair of VAST events. Complicating this interpretation are other effects that potentially caused appreciable temporal spreading of the arrivals. These effects include the extended nature (both in time and space) of the source, appreciable coupling of both shear wave and compressional wave energy into the water column (TALANDIER and OKAL, 1998), and dispersive propagation along the initial land path.

The ocean dispersion characteristics are strongly affected by quite short sections of shallow water along the propagation path. For example, the temporal spreading of nearly 7 s at the higher frequencies due to propagation from Loihi to the base of the continental slope off the west coast of the U.S. is decreased by about 2 s by the additional 35-km path up the slope to the Pt. Sur station location. This reversal of the deep water dispersion reaches its extreme when the water becomes so shallow that coupling into land vibrations occurs. The scrambling of the relative timing between the normal modes is strongly dependent upon the details of the ocean bottom bathymetry near the places of coupling. This scrambling is illustrated by the recordings of two Hawaiian events by the *T*-phase station VIB and the land seismic station BKS, which display significant differences in arrival structure.

Acknowledgements

We would like to thank Keith McLaughlin, now at the Center for Monitoring Research, who was instrumental in initiating this project. Thanks also to Marcia McLaren of Pacific Gas and Electric for the use of her seismic data and for pointing out to us the large *T* phases from the Hawaiian and Loihi events. Leland Treebs of MPL scanned the LVLA data from the VAST experiment for significant *T*-phase arrivals. Beneficial discussions have been held with Paul Hursky, Jim Murray, and Bill Hodgkiss of MPL, Mariana Eneva of Maxwell Technologies, Jackie Caplan-Auerbach at the School of Ocean and Earth Science and Technology, Peter Gerstoft at the Comprehensive Nuclear-Test-Ban Treaty Organization in Vienna, Austria, and Catherine de Groot-Hedlin at the Institute for Geophysics and Planetary Physics, Scripps Institution of Oceanography. Helpful comments were made by one of the paper's reviewers. John Hildebrand of MPL was responsible for the design of the LVLA array and its deployment during the 1989 VAST experiment. Finally, Terry Barker formerly of Maxwell Technologies was one of the original contributors to this project.

Many of the results in this paper were presented in the special session, "Acoustics and the Comprehensive Test-Ban Treaty," at the Spring, 1999, Acoustical Society Meeting in Berlin, Germany (D'SPAIN *et al.*, 1999).

This research was supported by the Defense Threat Reduction Agency under contract #DSWA01-97-C-0166. The construction of the LVLA array and data collection during the VAST experiment was sponsored by the Office of Naval Research.

Appendix

Spectral Ratio Grams for the VAST Events

The calibrated spectrograms for the two VAST events are presented in Section I (Figs. 3 and 4). In contrast, the spectral ratio spectrograms for the Hawaiian and

French nuclear test events are given in Section II (Figs. 16, 17, 20, 21, and 22). To make the comparison between these two sets of events easier for the reader, the spectral ratio spectrograms for the two VAST earthquakes over a 2-min. period and from 0–30 Hz are plotted in Figures A1 and A2. The VAST events display a more complicated single-element arrival structure than the Hawaiian and Mururoa events. The cause of the complicated appearance is probably due to more complex source time functions for these events, although differences in solid-earth-to-water coupling off California and off the mid-ocean islands of Hawaii and Mururoa may also play an important role.

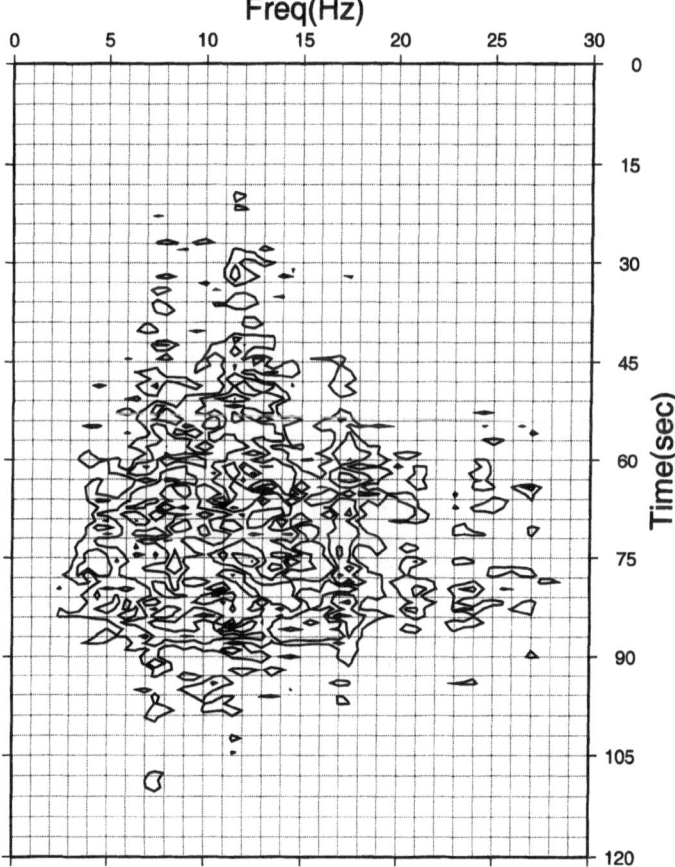

Figure A1

Contour plot of the spectral ratio spectrogram for VAST event 3, an m_b 4.1 earthquake offshore at Mendocino Ridge (Fig. 1 and Table 1 in the text). The corresponding calibrated spectrogram for this event is shown in Figure 3. The contours are plotted in 6 dB steps from 22 dB to 46 dB.

Freq(Hz)

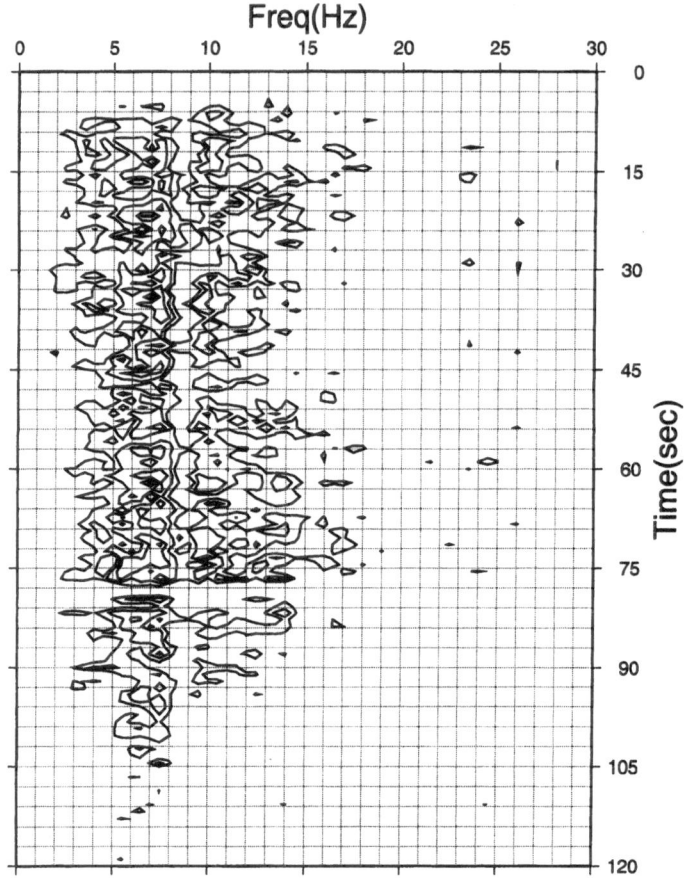

Figure A2

Spectral ratio spectrogram contour plot for VAST event 4, the earthquake on land near San Jose (Fig. 1 and Table 1 in the text). Its calibrated spectrogram appears in Figure 4. The contours are plotted in 6 dB steps from 10 dB to 34 dB.

REFERENCES

BURENKOV, S. V. (1989), *Distinctive Features of the Interference Structure of a Sound Field in a Two-dimensionally Inhomogeneous Waveguide*, Sov. Phys. Acoust. *35*(5), 465–467 (Akust. Zh. 35, 797–800).

BYERLY, P., and Herrick, C. (1954), *T Phases from Hawaiian Earthquakes*, Seis. Soc. Am. Bull. *44*(2a), 113–121.

CAPLAN-AUERBACH, J., DUENNEBIER, F. K., OKUBO, P., and KONG, L. (1996), *Characteristics of the 1996 Loihi Seismic Swarm*, EOS Transact., Am. Geophys. Union Fall Meeting Suppl.

CAPLAN-AUERBACH, J., and DUENNEBIER, F. (2000), *Seismicity and Velocity Structure of Loihi Seamount from the 1996 Earthquake Swarm*, submitted to Seis. Soc. Am. Bull.

COLLINS, M. D. (1995), *Beach Acoustics*, J. Acoust. Soc. Am. *97*(5), pt. 1, 2767–2770.

COMPREHENSIVE NUCLEAR-TEST-BAN TREATY ORGANIZATION (1998), Website, http://www.ctbto.org/ctbto.

COPPENS, A. B., and SANDERS, J. V., *Transmission of Sound into a Fast Fluid Bottom from an Overlying Fluid Wedge*. In *Proceedings of Workshop on Seismic Propagation in Shallow Water* (Office of Naval Research, Arlington, VA, 1978).

D'SPAIN, G. L., HODGKISS, W. S., and EDMONDS, G. L. (1991), *Energetics of the Deep Ocean's Infrasonic Sound Field*, J. Acoust. Soc. Am. *89*(3), 1134–1158.

D'SPAIN, G. L., BERGER, L. P., KUPERMAN, W. A., STEVENS, J. L., and BAKER, G. E. (1999), *Normal Mode Composition of Earthquake T Phases Recorded in the Deep Ocean*, J. Acoust. Soc. Am. *105*(2), pt. 2, 1039.

D'SPAIN, G. L., and KUPERMAN, W. A. (1999), *Application of Waveguide Invariants to Analysis of Spectrograms from Shallow Water Environments that Vary in Range and Azimuth*, J. Acoust. Soc. Am. *106*(5), 2454–2468.

D'SPAIN, G. L., BERGER, L. P., KUPERMAN, W. A., HODGKISS, W. S., DORMAN, L. M., and GAINES, W. A. (2000), *Observations of Land Vehicle Activity and Other Land-based Sounds with Offshore Underwater Acoustic Arrays*, J. Acoust. Soc. Am., in press.

DUENNEBIER, F. K., BECKER, N. C., CAPLAN-AUERBACH, J., CLAGUE, D. A., COWEN, J., CREMER, M., GARCIA, M. O., GOFF, E., MALAHOFF, A., McMURTRY, G. M., MIDSON, B. P., MOYER, C. L., NORMAN, M., OKUBO, P., RESING, J. A., RHODES, J. M., RUBIN, K., SANSONE, F. J., SMITH, J. R., SPENCER, K., WEN, XIYUAN, and WHEAT, C. G. (1997), *Researchers Rapidly Respond to Submarine Activity at Loihi Volcano, Hawaii*, EOS Transact., Am. Geophys. Union *78*(22), 229–233.

EWING, W. M., and WORZEL, J. L. (1948), *Long-range Sound Transmission*, Geo. Soc. Am. Memoir *27*.

GRINDA, L. (1960), *Nouveaux aspects des ondes T (New Aspects of T waves)*, Comp. Rend. Acad. Sci. *250*(12), 2241–2243.

DE GROOT-HEDLIN, C., *Computation of T-phase Coda*. In *Proceedings 16th International Congress on Acoustics and 135th Meeting of Acoust. Soc. Am.*, vol. 1 (Acoustical Society of America, New York, 1998) pp. 25–26.

DE GROOT-HEDLIN, C., and ORCUTT, J. A. (1999), *Synthesis of Earthquake-generated T Waves*, Geophys. Res. Lett. *26*(9), 1227–1230.

HAWAII CENTER FOR VOLCANOLOGY, School of Ocean and Earth Science and Technology, University of Hawaii (1998), Website, http://www.soest.hawaii.edu.

HEANEY, K. D., and KUPERMAN, W. A. (1998), *Very Long-range Source Localization with a Small Vertical Array*, J. Acoust. Soc. Am. *104*(4), 2149–2159.

HODGKISS, W. S., *Downslope Conversion*. In *Acoustic Signal Processing for Ocean Exploration* (ed. Moura, J. M. F., and Lourtie, I. M. G.) (Kluwer Academic Publ., Netherlands, 1993) pp. 145–150.

HOWE, B. M., MERCER, J. M., SPINDEL, R. C., WORCESTER, P. F., HILDEBRAND, J. A., HODGKISS, W. S., DUDA, T. F., and FLATTÉ, S. M., *SLICE 89: A Single Slice Tomography Experiment*. In *Proceedings of the Workshop on Ocean Variability and Acoustic Propagation* (eds. Potter, J., and Warn-Varnas, A.) (Kluwer Acad. Publ., The Netherlands, 1991) pp. 81–86.

HURSKY, P., HODGKISS, W. S., and KUPERMAN, W. A. (1995), *Extracting Modal Structure from Vertical Array Ambient Noise Data in Shallow Water*, J. Acoust. Soc. Am. *98*, pt. 2, 2971.

HURSKY, P., HODGKISS, W. S., and KUPERMAN, W. A. (2000), *Matched Field Processing with Data-derived Modes*, submitted to J. Acoust. Soc. Am.

JENSEN, F. B., and KUPERMAN, W. A. (1980), *Sound Propagation in a Wedge-shaped Ocean with a Penetrable Bottom*, J. Acoust. Soc. Am. *67*(5), 1564–1566.

JENSEN, F. B., KUPERMAN, W. A., PORTER, M. B., and SCHMIDT, H., *Computational Ocean Acoustics* (American Institute of Physics Press, New York, 1994).

JOHNSON, R. H. (1963), *Spectrum and Dispersion of Pacific T Phases*, Hawaii Inst. Geophys. Rept. 34, Tech. Summary Rept. 4, University of Hawaii, Honolulu, Hawaii.

JOHNSON, R. H., NORTHROP, J., and EPPLEY, R. (1963), *Sources of Pacific T Phases*, J. Geophys. Res. *68*(14), 4251–4260.

KUPERMAN, W. A., and INGENITO, F. (1980), *Spatial Correlation of Surface Generated Noise in a Stratified Ocean*, J. Acoust. Soc. Am. *67*(6), 1988–1996.

KUPERMAN, W. A., D'SPAIN, G. L., and HEANEY, K. D. (2000), *Long-range Source Localization from Single Hydrophone Spectrograms*, accepted for publ. in J. Acoust. Soc. Am.

MILNE, A. R. (1959), *Comparison of Spectra of an Earthquake T Phase with Similar Signals from Nuclear Explosions*, Seis. Soc. Am. Bull. *49*, 317–330.

NATIONAL EARTHQUAKE INFORMATION CENTER, U.S. Geological Survey (1998), Website, http://www.neic.cr.usgs.gov/neis/epic/epic.html.

NATIONAL GEOPHYSICAL DATA CENTER, National Oceanic and Atmos. Admin. (1998), 5-Min Global DTM, Website, http://www.ngdc.noaa.gov/seg/fliers/se-1104.html.

NEILSEN, T. B., WESTWOOD, E. K., and UDAGAWA, T. (1997), *Mode Function Extraction from a VLA Using Singular Value Decomposition*, J. Acoust. Soc. Am. *101*(5), pt. 2, 3025.

NEMO OCEANOGRAPHIC DATA SERVER, Scripps Institution of Oceangraphy (1998), National Oceanographic Data Center's Station Data File, Website, http://nemo.ucsd.edu/hs.html.

NOBLE, B., *Applied Linear Algebra* (Prentice-Hall, Inc, New Jersey, 1969).

NORTHROP, J. (1962), *Evidence of Dispersion in Earthquake T Phases*, J. Geophys. Res. *67*(7), 2823–2830.

NORTHROP, J. (1974), *T Phases from the Hawaiian Earthquake of April 26, 1973*, J. Geophys. Res. *79*(35), 5478–5481.

NORTHROP, J., and JOHNSON, R. H. (1965), *Seismic Waves Recorded in the North Pacific from FLIP*, J. Geophys. Res. *70*(2), 311–318.

OLIVERA, M., MURRAY, J. J., BUONO, L., and HODGKISS, W. S. (1994), *Modal Decomposition: VAST Track W*, MPL Tech. Memo. 439, Marine Physical Laboratory of the Scripps Institution of Oceanography, San Diego, Calif.

PORTER, M. B. (1991), *The Kraken Normal Mode Program, User's Manual*, SACLANT Undersea Research Centre, La Spezia, Italy.

PROTOTYPE INTERNATIONAL DATA CENTER (1998), Website, http://www.pidc.org/dataprodbox/prod.html.

RANGE-DEPENDENT BENCHMARK PROBLEMS in OCEAN ACOUSTICS (1990), Special section of J. Acoust. Soc. Am. *87*(4), 1497–1545.

SCHMIDT, R. O. (1981), *A Signal Subspace Approach to Multiple Emitter Location and Spectral Estimation*, Ph.D. Thesis, Stanford Univ., Stanford, CA.

SMITH, G. B. (1997), *"Through the Sensor" Environmental Estimation*, J. Acoust. Soc. Am. *101*(5), pt. 2, 3046.

SOTIRIN, B. J., and HILDEBRAND, J. A. (1988), *Large Aperture Digital Acoustic Array*, IEEE J. Oceanic Engin. *13*(4), 271–281.

STEVENS, J. L., BAKER, G. E., MURPHY, J. R., COOK, R. W., D'SPAIN, G. L., BERGER, L. P., and KHRISTO-FOROV, B. D. (1998), *T-phase Excitation and Transfer Function Research*, Proc. 20th Seismic Research Symposium, 698–707.

STEVENS, J. L., BAKER, G. E., COOK, R. W., D'SPAIN, G. L., BERGER, L. P., and DAY, S. M. (2000), *Empirical and Numerical Modeling of T-phase Propagation from Ocean to Land*, Pure appl. geophys., in press.

TALANDIER, J., and OKAL, E. A. (1998), *On the Mechanism of Conversion of Seismic Waves to and from T Waves in the Vicinity of Island Shores*, Seis. Soc. Am. Bull. *88*(2), 621–632.

TINDLE, C. T., GUTHRIE, K. M., BOLD, G. E. J., JOHNS, M. D., JONES, D., DIXON, K. O., and BIRDSALL, T. G. (1978), *Measurements of the Frequency Dependence of Normal Modes*, J. Acoust. Soc. Am. *64*(4), 1178–1185.

TOLSTOY, I., and EWING, M. (1950), *The T Phase of Shallow-focus Earthquakes*, Seis. Soc. Am. Bull. *40*(1), 25–51.

WOLF, S. N. (1987), *Experimental Determination of Modal Depth Functions from Covariance Matrix Eigenfunction Analysis*, J. Acoust. Soc. Am. Suppl. 1., *81*(S1), 64.

WOLF, S. N., COOPER, D. K., and ORCHARD, B. J., *Environmentally Adaptive Signal Processing in Shallow Water*. In *Oceans '93, Engineering in Harmony with Ocean Proceedings*, Victoria, BC, Canada, vol. I, 1993, pp. 99–104.

(Received June 20, 1999, revised April 4, 2000, accepted April 5, 2000)

To access this journal online:
http://www.birkhauser.ch

Pure appl. geophys. 158 (2001) 513–530
0033–4553/01/030513–18 $ 1.50 + 0.20/0

| Pure and Applied Geophysics

T-phase Observations in Northern California: Acoustic to Seismic Coupling at a Weakly Elastic Boundary

CATHERINE DEGROOT-HEDLIN[1] and JOHN ORCUTT[1]

Abstract — Plans for a hydroacoustic network intended to monitor compliance with the CTBT call for the inclusion of five *T*-phase stations situated at optimal locations for the detection of seismic phases converted from ocean-borne *T* phases. We examine factors affecting the sensitivity of land-based stations to the seismic *T* phase. The acoustic to seismic coupling phenomenon is described by upslope propagation of an acoustic ray impinging at a sloping elastic wedge. We examine acoustic to seismic coupling characteristics for two cases; the first in which the shear velocity of the bottom is greater than the compressional velocity of the fluid (i.e., $v_p > v_s > v_w$), the second is a weakly elastic solid in which $v_s \ll v_w < v_p$. The former is representative of velocities in solid rock, which might be encountered at volcanic islands; the latter is representative of marine sediments. For the case where $v_s > v_w$, we show that acoustic energy couples primarily to shear wave energy, except at very high slope angles. We show that the weakly elastic solid (i.e., $v_s \ll v_w$) behaves nearly like a fluid bottom, with acoustic energy coupling to both *P* and *S* waves even at low slope angles.

We examine converted *T*-wave arrivals at northern California seismic stations for two event clusters; one a series of earthquakes near the Hawaiian Islands, the other a series of nuclear tests conducted near the Tuamoto archipelago. Each cluster yielded characteristic arrivals at each station which were consistent from event to event within a cluster, but differed between clusters. The seismic *T*-phases consisted of both *P*- and *S*-wave arrivals, consistent with the conversion of acoustic to seismic energy at a gently sloping sediment-covered seafloor. In general, the amplitudes of the seismic *T* phases were highest for stations nearest the continental slope, where seafloor slopes are greatest, however noise levels decrease rapidly with increasing distance from the coastline, so that *T*-wave arrivals were observable at distances reaching several hundred kilometers from the coast. Signal-to-noise levels at the seismic stations are lower over the entire frequency spectrum than at the Pt. Sur hydrophone nearby, and decrease more rapidly with increasing frequency, particularly for stations furthest from the continental slope.

Key words: Hydroacoustics, *T* phase, acoustic-to-seismic conversion.

Introduction

The planned hydroacoustic network designed to monitor compliance with the Comprehensive Test-Ban Treaty (CTBT) will eventually consist of four hydrophones and five *T*-phase stations located on ocean islands. The plan is to locate *T*-phase stations, each consisting of a single vertical component seismometer, at optimal sites

[1] Scripps Institution of Oceanography, La Jolla, CA 92093-0225, USA. E-mail: cdh@eos.ucsd.edu, jorcutt@igpp.ucsd.edu

for detecting seismic phases generated by the coupling of acoustic T phases into compressional (P) or shear (S) waves. Since seismic phases are more severely attenuated than ocean-borne acoustic energy, the T-phase stations are expected to be less sensitive to acoustic energy excited by nuclear tests and low atmospheric explosions than hydrophones are. However, their anticipated lower cost makes their use desirable.

Although T phases have been detected at stations with long inland paths (e.g., TOLSTOY and EWING, 1950; CANSI and BETHOUX, 1985), the rationale behind locating T-phase sensors close to a shoreline is that T waves are expected to undergo less attenuation, particularly at high frequencies, at these locations. The ability to detect T waves over a broad frequency band is an essential feature of a T phase station, as naturally-occurring events often have different spectral characteristics than explosions. Suboceanic earthquakes usually generate low frequency acoustic phases that peak at about 5 Hz. Underwater explosions yield a characteristic bubble pulse (COLE, 1948) which is manifested in the frequency domain as spectral scalloping.

In this paper, we examine factors affecting the sensitivity of land-based stations to the seismic T phase. In the next section, we examine the physics of acoustic to seismic coupling at a sloping wedge for an acoustic bottom and for two types of elastic bottoms; a Poisson solid with a shear velocity (v_s) that is higher than the acoustic velocity (v_w) of the fluid above, and a semi-elastic solid with a shear velocity considerably lower than v_w. In the following section, we examine T-phase arrivals at seismic stations in northern California for two event clusters. The first is a series of shallow events that occurred near the Hawaiian Islands in July 1996; the second is a series of nuclear tests conducted by the French in the Tuamoto archipelago from September 1995 to January 1996. Both types of events generated T-waves that were recorded at a Pt. Sur hydrophone as high amplitude, short duration (15–20 sec), broadband arrivals. However, each type of event yielded characteristic T-phase recordings at the seismic stations that were repeatable within each cluster, but distinct between clusters.

T-phase Coupling at an Elastic Bottom

Long-range acoustic propagation in the SOFAR channel can be described in terms of adiabatic normal mode theory (e.g., HEANEY et al., 1991). At any given frequency, a mode may be represented by an equivalent pair of upgoing and downgoing rays that propagate at an angle or phase velocity giving rise to constructive interference (OFFICER, 1958). The propagation angle is constant for a range-independent ocean, with higher modes propagating at steeper angles than low modes. As the acoustic wavefield approaches a sufficiently shallow, upward sloping seafloor, a ray trapped in the SOFAR channel becomes bottom interacting, and steepens by twice the seafloor inclination at each cycle of bottom and surface reflections. As the angle becomes steeper than the critical angle, part of the energy is transmitted into the seafloor at each successive bottom bounce. Parabolic equation modeling indicates that, for a fluid over a fluid wedge, acoustic energy is radiated into the bottom in discrete beams at approximately the depths predicted by normal-mode theory (JENSEN and KUPERMAN, 1980). Thus, for a fluid bottom, the acoustic T phase would couple into P energy at discrete segments along the slope, defined by the cutoff depths.

The mode cutoff depth is dependent on both mode number and frequency. At any given frequency, low modes have the shallowest cutoff depth; for any given mode, the highest frequencies have the shallowest cutoff depth. The lowest modes and highest frequencies are therefore predicted to have the longest oceanic travel path and to couple into seismic energy nearest the shore. The depth dependence of the acoustic to seismic coupling thus translates to a time separation between coupling points. However, separate T-wave arrivals corresponding to individual mode cutoffs are expected only for highly impulsive sources. In general, for shallow slopes, seismic T phases with longer duration than the incident acoustic phase are predicted due to the greater lateral separation between the mode cutoff depths. For steeper slopes, mode cutoff depths are more closely spaced, so the seismic T phase would be expected to have a greater impulsive arrival, yielding T phases with greater SNR than for shallow slopes. In fact, this behavior has been observed in several studies (e.g., TALANDIER and OKAL, 1998; HANSON, 1998); that is, efficient T-phase conversion from T waves in the water column to seismic phases occurs in the presence of steep slopes.

Thus far we have discussed only the fluid over fluid case, where the S-wave velocity in the bottom is assumed to be zero. If instead we consider an elastic bottom, with non-zero shear velocity, the coupling problem becomes more complicated. The ray propagation scenario described above is modified by the fact that there are now two critical angles, θ_s, corresponding to the critical angle for transmission of shear waves into the bottom, and θ_p corresponding to transmission of compressional waves. Since P velocities are always greater than S velocities by at least a factor of $\sqrt{2}$, the critical angle for S waves is less steep than that for P waves. Therefore, ray theory predicts that supercritical incidence is reached first for the S wave, then for the P wave after several more bottom reflections.

The partitioning of energy into P and S waves is strongly dependent upon both the slope and the velocity structure of the elastic bottom. Energy flux densities for transmission and reflection of an acoustic wave incident on an elastic solid, computed using equations (4.2.24–4.2.36) of BREKHOVSKIKH and GODIN (1990), are shown in Figure 1 at a range of incident angles for three elastic wedge models. The velocities and densities, listed in the figure caption, are typical for solid rock near the earth's surface. For a ray incident on a shallow elastic wedge, the propagation angle with respect to the vertical decreases by twice the slope inclination at each bottom bounce until it falls below critical incidence for the shear wave in the bottom. A portion of the acoustic energy then couples into S-wave energy in the solid at each bottom

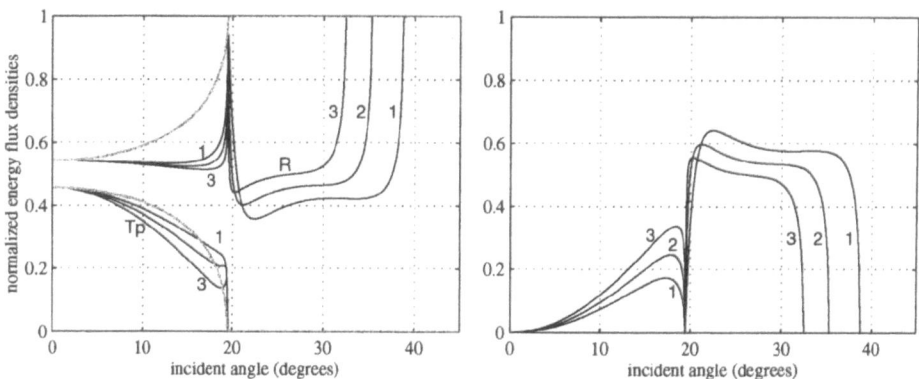

Figure 1
Energy flux densities for transmission and reflection of an acoustic ray incident on an elastic solid. The
ratio of compressional velocities from solid to fluid is 3:1, corresponding to a P-wave velocity of 4.5 km/sec
and a water velocity of 1.5 km/sec. The density contrast from solid to liquid is 2.2. Curves were computed
for three values of S-wave velocity: (1) 2.4 km/sec, (2) 2.6 km/sec (corresponding to a Poisson solid), and
(3) 2.8 km/sec. At left are shown the reflected energy fluxes (R) and the transmitted P-wave energy (Tp).
Corresponding values for an acoustic ray incident on an equivalent liquid, i.e., one with identical
compressional velocity and density are shown by the gray lines. Transmitted S-wave energy flux densities
are shown in the right panel. Velocities and densities for the elastic solid are typical for those of solid rock.

bounce until the critical angle for the P wave is reached. Any remaining acoustic
energy then couples into both the P and S waves as the angle of incidence continues
to decrease. Overall, the introduction of a non-zero shear velocity "softens" the
boundary, that is, the total impedance of the solid is less than that of an equivalent
liquid with identical velocity and density values, thus the reflection coefficient in the
liquid is reduced (BREKHOVSKIKH and LYSANOV, 1991), as indicated by the gray lines
which show reflection and transmission coefficients for a liquid with identical density
and compressional velocities. For $\theta < \theta_p$, the reflection and transmission coefficients
vary more with angle of incidence than for the equivalent solid.

For a slope with inclination α, a ray goes through approximately $(\theta_s - \theta_p)/2\alpha$
bottom bounces between reaching the critical angles for S and P waves. Part of the
acoustic energy is transformed into shear wave energy in the bottom at each bounce,
thus a shallower slope results in more bounces during which acoustic energy can be
transformed to shear waves. This may account for the observations of TALANDIER
and OKAL (1998) that for Polynesian shield islands (i.e., regions where the seafloor is
composed of recent volcanics or coral reefs) $T \rightarrow P$ waves are the dominant arrivals
only at very steep slopes (i.e., slopes greater than 50°), and that the energy couples
instead to shear waves at 20° slopes. Also, parabolic equation modeling confirms that
acoustic energy couples into S-wave energy for shallow slopes (COLLINS, 1993), for
the case in which the bottom shear velocity is greater than the water velocity. As
indicated in Figure 1, the transmission factor for the S waves increases as shear
velocity decreases (for constant P velocity), and conversely, the amount of energy

reflected at each bounce decreases. Furthermore, $\theta_s - \theta_p$ increases for decreasing shear velocity so that the range of incident angles over which acoustic energy is transmitted into S waves increases. Thus, for this range of velocities and densities, the percent of energy coupling into S waves increases with decreasing shear velocities and decreasing slope.

For typical marine sediments, the shear velocity of the elastic bottom is less than the compressional velocity of the fluid above (HAMILTON, 1980), so that an incident T phase is post-critical with respect to the bottom S waves at all angles. Thus $T \rightarrow S$ conversion takes place whenever the seafloor intersects the SOFAR velocity channel. Energy flux densities for transmission and reflection of an acoustic wave incident on an elastic solid with velocities and densities corresponding to those of marine sediments are shown in Figure 2. As shown, for angles of incidence θ such that $\theta_p < \theta < \theta_s$, the transmission factors for S waves increase with increasing shear velocity, with most of the energy reflected at each bottom bounce. Once θ_p is reached, most of the energy is transmitted into P waves thus, in this case, the percentage of energy coupling into S waves increases with increasing shear velocity. Again, we predict that the relative partitioning of energy between P and S waves will depend on the slope, with coupling into shear waves increasing with decreasing slope. For a shear velocity of zero, the problem reduces to acoustic transmission only and the acoustic energy couples into P waves in the bottom.

Transmission and reflection coefficients for an equivalent liquid bottom are plotted in gray in Figure 2. As indicated, the reflection and transmission coefficients for $\theta < \theta_p$ are almost identical to those of the typical marine sediments, thus for low angles of incidence, the physics of $T \rightarrow P$ coupling are almost identical to that of the

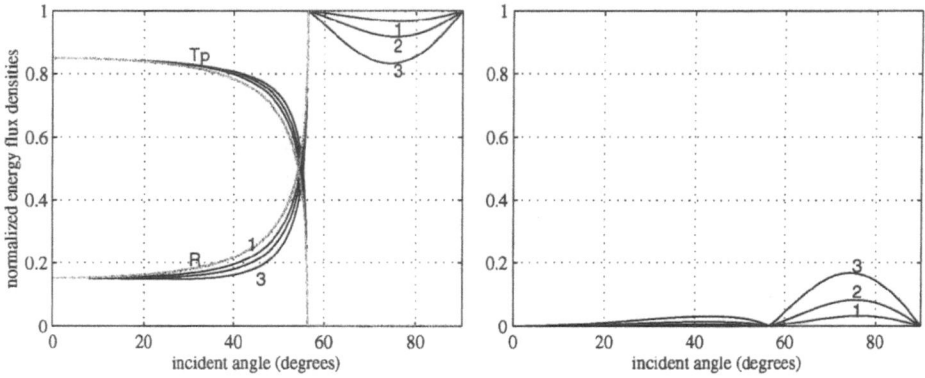

Figure 2

As for Figure 1, but for an elastic solid with velocities and densities corresponding to those of marine sediments. The P-wave velocity of the solid is 1.8 km/sec, the density contrast from solid to liquid is 1.9. Curves were computed for three values of S-wave velocity: (1) 0.3 km/sec, (2) 0.4 km/sec, and (3) 0.5 km/sec.

fluid over fluid wedge. As shown by JENSEN and KUPERMAN (1980), for a fluid over fluid wedge with low velocity contrast, the water-borne energy is radiated into downward-going P waves in the bottom at depth ranges corresponding to mode cutoff depths (JENSEN and KUPERMAN, 1993). This suggests that there may be a shadow zone for $T \rightarrow P$ coupling, with the extent of the shadow zone being dependent on the P-velocity gradient. Since shear velocities in the bottom are considerably lower than the water velocity, acoustic energy coupled into shear energy will also be refracted downward into the bottom, creating a $T \rightarrow S$ shadow zone. Hence $T \rightarrow S$ or $T \rightarrow P$ phases may be absent for stations located near the coast.

Comparison of Acoustic and Seismic T-phase Recordings

A map of the seismic stations used in this study is presented in Figure 3, along with the near-shore bathymetry in the region. Also shown are the relative positions of the Hawaiian and nuclear test events with respect to the receivers. The seismic stations on land, which form part of the Berkeley seismic array, are equipped with

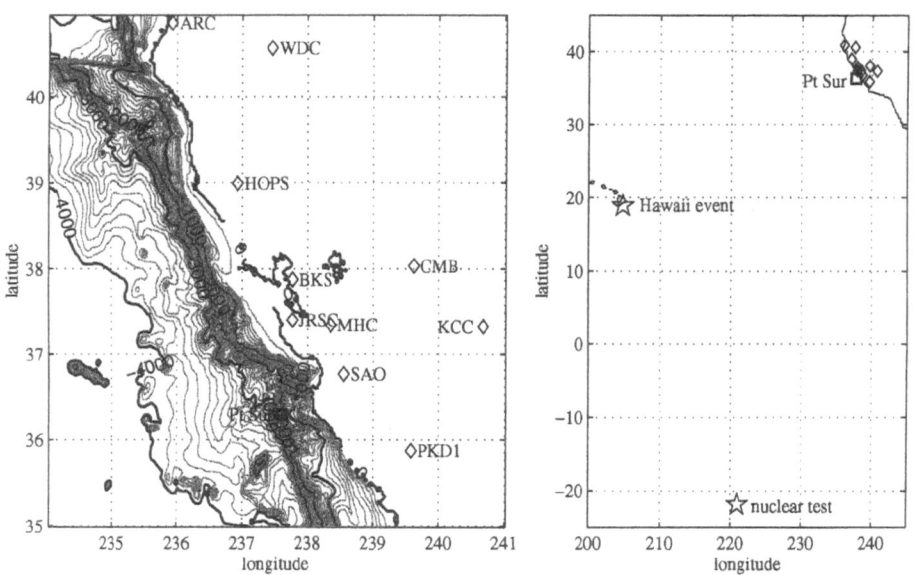

Figure 3
At left is shown a map of stations used in this study. The locations of the seismic stations, which form part of the Berkeley seismic array, are indicated by diamonds; the location of the Pt. Sur hydrophone is marked by a square. Contour lines delineate the bathymetry (from SMITH and SANDWELL, 1997) near the shoreline. The heavy lines show contours every 1 km from 0 to 4 km depth; the thinner grey lines show contours every 250 m. The locations of the French nuclear test and the Hawaii earthquake relative to the Pt. Sur hydrophone and the Berkeley seismic array are shown at right.

three-component accelerometers with a 20 Hz sampling rate. The hydrophone at Pt. Sur measures pressure at a 200 Hz sampling rate. The continental slope in this region is fairly gradual, with average slopes of about 5° between 200 m and 3.2 km depth. The seafloor off California consists of enormous coalescing fans (the Delgada Fan and the Monterey Fan) that cover the continental shelf and slope, and transgress over the abyssal hills province (SEIBOLD and BERGER, 1996). Reflection profiles and drill logs recovered at DSDP sites 32 (McMANUS et al., 1970), and 33 (McMANUS et al., 1970) in central California, and 467 (YEATS et al., 1981) in southern California, indicate that the sediment column is several hundreds of meters thick in the deep water of the continental rise but is well over 1 km thick on the continental slope. Compressional velocities in these marine sediments range from 1.5–1.6 km/sec in the top 500 m, increasing to 3 km/sec at a depth of 1 km, and densities grade from 1600 kg/m^3 to 2300 kg/m^3. Waveform fitting of interface waves indicates that shear velocities range from 30 m/sec at the surface to 450 m/sec at a depth of 150 m (NOLET and DORMAN, 1996).

Vertical component recordings of the seismic T waves, high-pass filtered above 2 Hz, are shown in the top panel of Figure 4 for a magnitude M_b 5.0 event, one of the largest of the Hawaiian event clusters of July-August, 1996. The characteristics of these arrivals, i.e., duration, frequency content, and relative amplitudes, are nearly uniform for events within the cluster. The fourth through tenth recordings are shown on a uniform physical scale, with the top three recordings, from stations further inland (WDC, CMB, and KCC), shown magnified ten times with respect to the others. The 'x' below each trace marks the predicted arrival time for entirely oceanic propagation along a geodesic path from source to receiver. Since seismic phases are faster than the acoustic T phase, which has a velocity of about 1.5 km/sec, all seismic T-phase arrivals are expected to precede this time if acoustic to seismic coupling takes place along the geodesic path. The Pt. Sur recording is shown for comparison in the bottom panel.

The signal-to-noise ratios (SNRs), computed for each component at each seismic station, are compared to those at Pt. Sur in Figure 5. All seismic records reveal lower SNR than for the hydrophone at all frequencies. T phases recorded at Pt. Sur for the Hawaii cluster peak at 2–8 Hz, and drop off gradually to background levels at 60 Hz. The T waves at the seismic stations peak at about 2 Hz, then drop off more rapidly with increasing frequency. The low SNRs at seismic stations for frequencies less than 2 Hz are due to high noise in this frequency band. Since the sampling rate at the seismic stations is only 20 Hz, compared to 200 Hz at Pt. Sur, high frequencies cannot be examined, however for most stations the SNRs suggest that most seismic T phases decrease to background levels at 8 Hz. SAO is an exception, with a relatively slow decrease in power with increasing frequency. As expected, recordings at stations located over 100 km inland (WDC, CMB, and KCC) are considerably more attenuated than for stations close to the coast, and the dropoff in signal level with increasing frequency is greater. However, stations closest to the coast (JRSC, BKS,

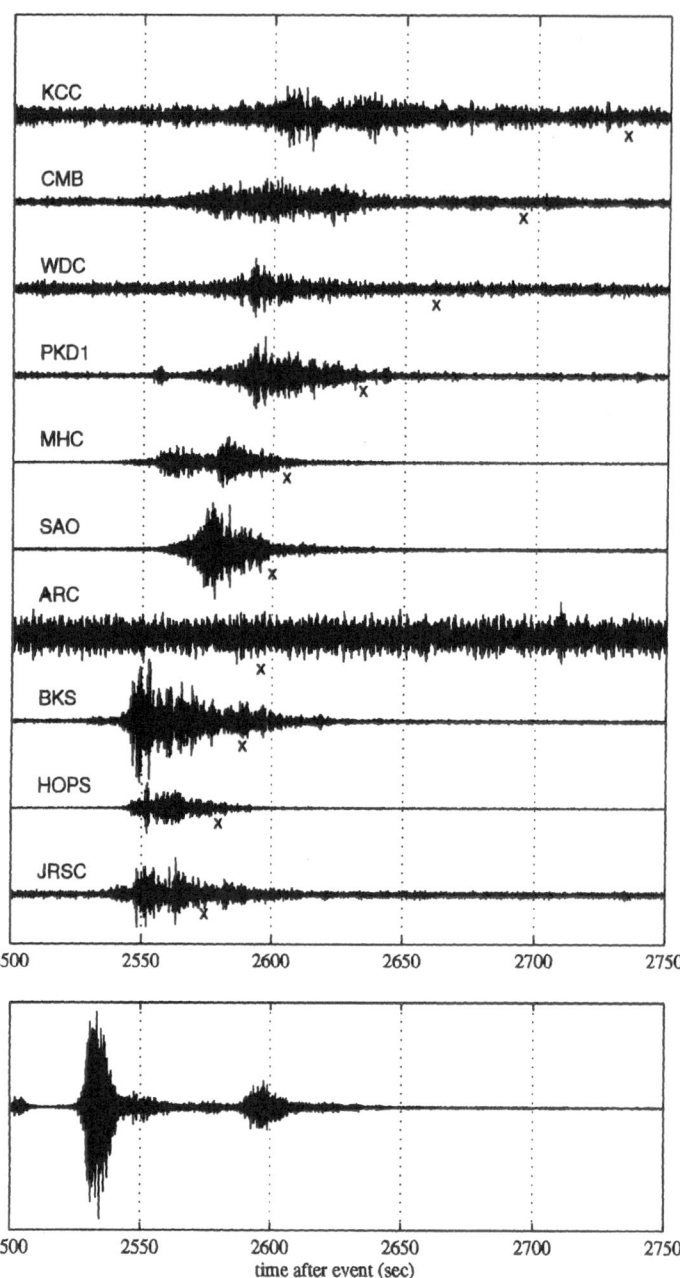

Figure 4

The top panel shows vertical records from the Berkeley array, high-pass filtered over 2 Hz, for a magnitude 5.0 event at 18.83, 155.32 W. The 'x' below each trace denotes the reference time corresponding to purely oceanic propagation from source to receiver. The fourth through tenth records are indicated at a uniform scale; the top three records, for the three stations furthest inland, are shown magnified ten times with respect to the others. For comparison the pressure data from the Pt. Sur hydrophone, bandpass-filtered from 2–10 Hz, is shown in the bottom panel.

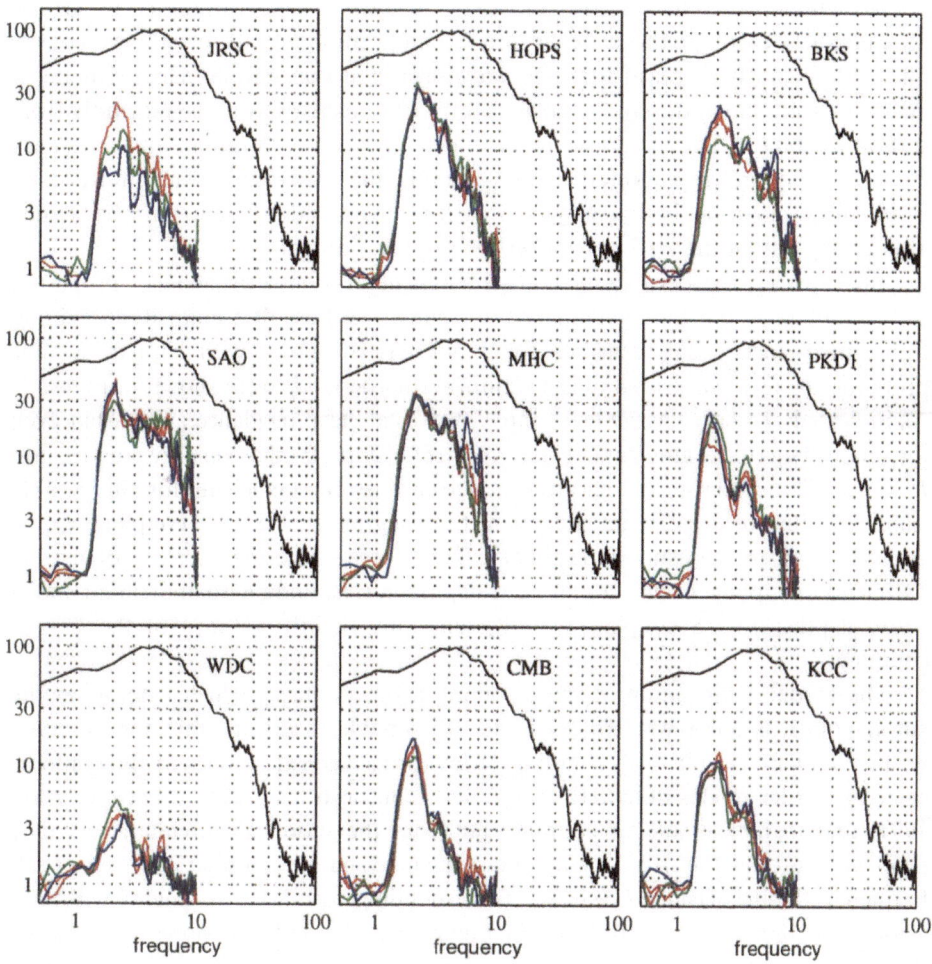

Figure 5

Signal-to-noise ratios (SNRs) are computed for each station. Ratios are computed by dividing the mean power spectrum over the entire *T*-phase arrival by the mean noise power for a quiet interval prior to the arrival. SNRs for the East-West (red), North-South (green), and vertical components (blue) are shown, compared to the SNR at Pt. Sur (black).

ARC) do not necessarily exhibit the highest SNR. This is due, in part, to higher noise levels at stations nearest the coastline, but is also due to the relative efficiency of acoustic to seismic coupling near each station, as we discuss next.

The signal characteristics of the seismic *T* phase are a result of the interaction of the acoustic wavefield with the elastic continental slope. Accurate prediction of the amplitude, duration, frequencies, and variations in $T \rightarrow P$ and $T \rightarrow S$ coupling would require an accurate model of seafloor velocities and bathymetry, as well as the relative mode strengths of the incoming T phase, and variations in crustal attenuation. Although the acoustic mode composition of the incoming acoustic phase is obviously identical for each waveform, and the bottom velocity model is nearly uniform along the coast, the waveform characteristics vary significantly from station to station. The variations in waveform amplitude, duration, and frequency content, which are repeatable for events within the Hawaiian cluster, must result from variations in the bathymetry near each receiver, and differences in the seismic attenuation within the land portion of the propagation path.

Since the shear velocity of the marine sediments is so low, the acoustic properties of the seafloor at the continental slope are similar to those of a fluid. Reflected and transmitted energy flux densities for a velocity structure with similar average properties to those measured in the California borderlands were shown in Figure 2. As indicated, T-phase energy leaks into S waves at steep incidence angles, but even for relatively gradual gradients at the continental slope, T-phase energy couples into P waves. The similarity of reflection and transmission coefficients for this weakly elastic medium to those of an equivalent fluid suggest that $T \rightarrow P$ coupling can be approximated by acoustic mode theory. Thus for a seafloor consisting of marine sediments, coupling occurs at discrete depths corresponding to individual mode cutoff depths; $T \rightarrow S$ coupling would occur only slightly seaward of the $T \rightarrow P$ coupling.

The modal composition of the incoming acoustic phase, which determines the cutoff points, is unknown. However, it is reasonable to assume that the observed seismic T phases result from coupling of low-order acoustic modes at the seafloor, since low-order modes dominate long-range acoustic transmission (HEANEY et al., 1991) and high-order modes coupling far from the shore are attenuated as they propagate through the crust. A map indicating mode cutoff depths at 2 Hz – the peak frequency for the seismic T waves – corresponding to marine sediments with a P velocity of 1800 m/sec is displayed in Figure 6 along with the slope angle in the direction of propagation from Hawaii to California. The cutoff depths for the lowest five modes occur along the continental slope, where the slopes are steepest, which suggests that the observed seismic T phases result from the coupling of the first five acoustic modes to seismic energy.

Also shown in Figure 6 are geodesic paths from source to receiver, for stations closest to the coastline. For the Hawaiian events, the direction of propagation is almost perpendicular to the continental slope, which implies that the geodesic travel path is a good approximation to the propagation path, since no lateral deflection takes place in passing from one medium to another at normal incidence. T-phase amplitudes are predicted to be larger for arrivals at direct incidence to the continental slope as compared to those at oblique incidence for three reasons. First, the seafloor slope is a maximum at direct incidence, therefore, as described in the previous section, converted modes arrive nearly concurrently, yielding high amplitudes. Secondly, since the acoustic wavefront is nearly parallel to the continental slope at

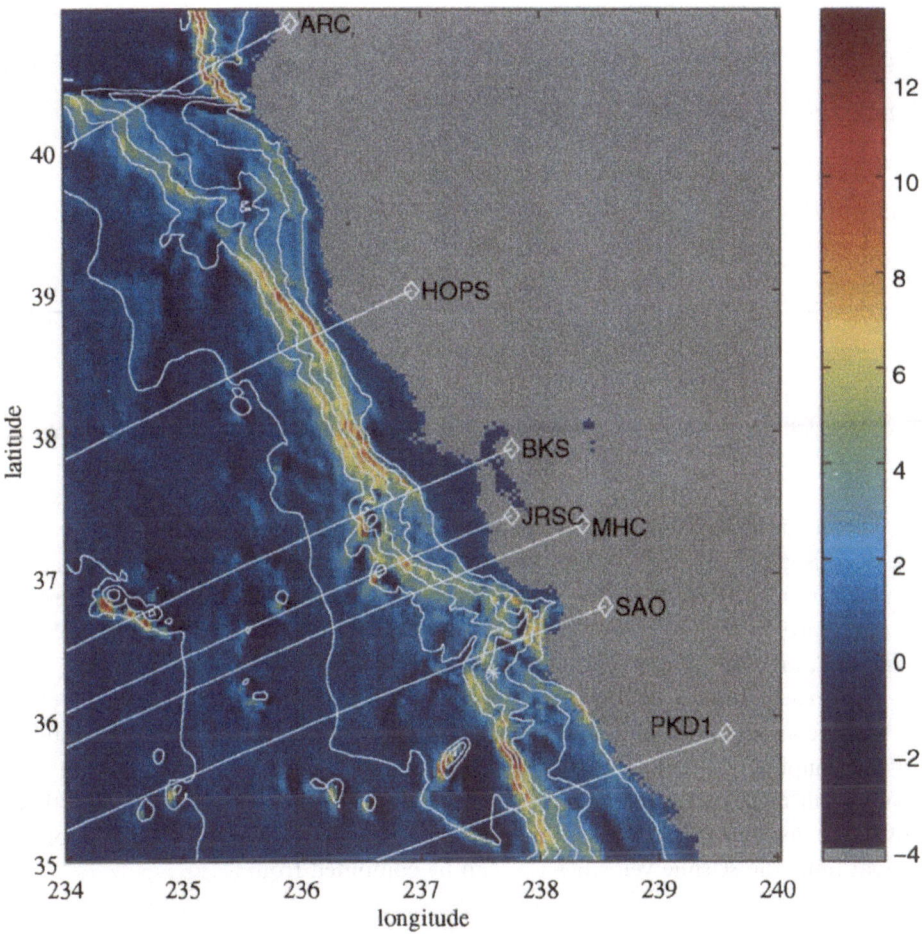

Figure 6

Slope angle (in degrees) in the direction of propagation from Hawaii to the Berkeley array, computed by taking the dot product of the gradients in bathymetry with the direction of propagation at each point. Contour depths corresponding to f = 2Hz cutoff depths for the first 7 modes are also shown for an acoustic seafloor with a compressional velocity of 1.8 km/sec. Geodesic paths from each station to the epicentral locations are also shown.

direct incidence, arrivals scattered from slightly off-axis also reach the receiver nearly concurrently, making the T phase relatively impulsive. Finally, angles of incidence at the seafloor are lower, on average, for arrivals normal to the coast and therefore correspond to higher transmission coefficients (cf. Figures 1 and 2). In the limit of very high angles of incidence, or very large velocity contrasts, acoustic energy is entirely reflected at the coastline.

Approximate horizontal distance ranges to where conversion takes place are shown in Figure 7 as a function of average SNR over 0–10 Hz for each station, where

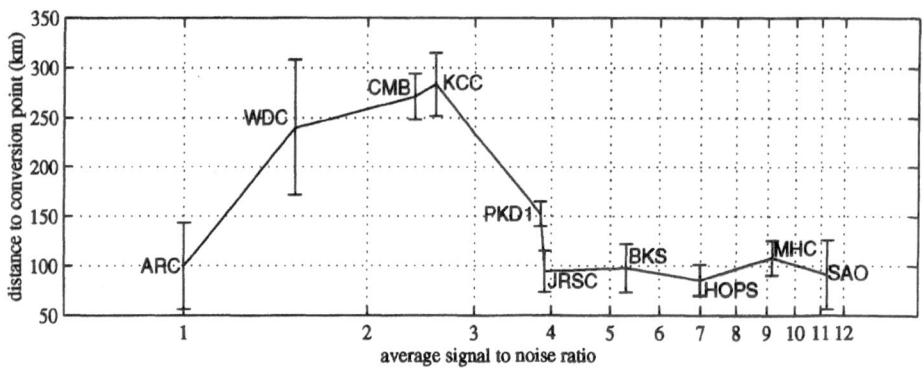

Figure 7
Lateral distance between the acoustic to seismic coupling point for each station as a function of the average
SNR over the 0–10 Hz band.

we have assumed that coupling takes place at ranges along the geodesic path where
the slope is greater than 3°. As expected, SNRs increase, on average, with decreasing
distance from the coupling point. However, as indicated by the point corresponding
to ARC, proximity to the coupling point does not ensure sensitivity to the T phase.
For ARC, high noise levels preclude its use as a T-phase station.

The greater duration of the recorded signals at the seismic stations (from 30 to 90
seconds), compared with that of the main T phase at Pt. Sur, suggests that acoustic to
seismic coupling takes place at different azimuthal ranges from each station, and may
result from coupling of T-phase energy into either S or P waves at the coastline.
Given the distance between the coupling point and receiver, and the seismic T-phase
arrival time, the seismic velocities v_{seis} can be computed from

$$dt = \Delta_{seis} \left(\frac{1}{v_w} - \frac{1}{v_{seis}} \right)$$

(as in TALANDIER and OKAL, 1998), where dt is the difference in travel time that of a
purely oceanic path from source to seismometer (marked by 'x' for each receiver in
Figure 4) and that of the seismic T-phase arrivals at each station, Δ_{seis} is the distance
travelled by the converted seismic phase (either P or S), and v_w is the velocity of the
acoustic phase in water. At JRSC and MHC which show two distinct arrivals, with
velocities of 4.0–5.7 km/sec for the first arrival, and 1.9–2.4 km/sec for the second
arrival, consistent with crustal P- and S-wave velocities, respectively. T phases at the
other stations appear to consist of several overlapping arrivals, with velocities
ranging from 5.0–6.2 km/sec for first arrivals, down to about 1.7 km/sec for the later
arrivals.

For comparison, we examine T phases recorded at the Berkeley array for one of a
series of nuclear tests conducted by the French in the Tuamoto Archipelago between
Sept. 5, 1995 and Jan. 27, 1996 (Fig. 8). The nuclear devices were buried to such a

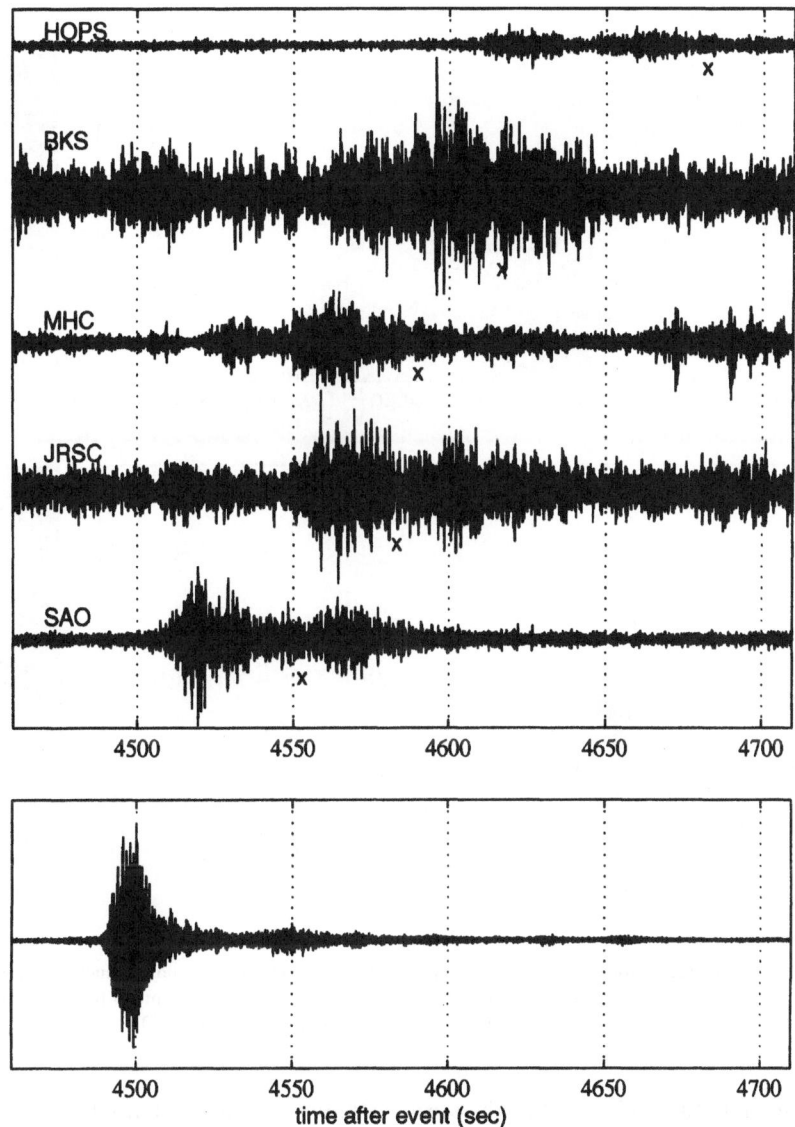

Figure 8
As in Figure 4, but for recording from the French nuclear test of October 27, 1995.

sufficient depth such that no shockwave entered the water (IDC Web Page), consequently the resulting T phase resembled that of a shallow earthquake. Spectral ratio sonograms at Pt. Sur for the Hawaiian event and the nuclear tests are compared in Figure 9. The duration of these acoustic phases is similar, although the dominant frequency of the nuclear test is markedly higher than that of the Hawaii event. The

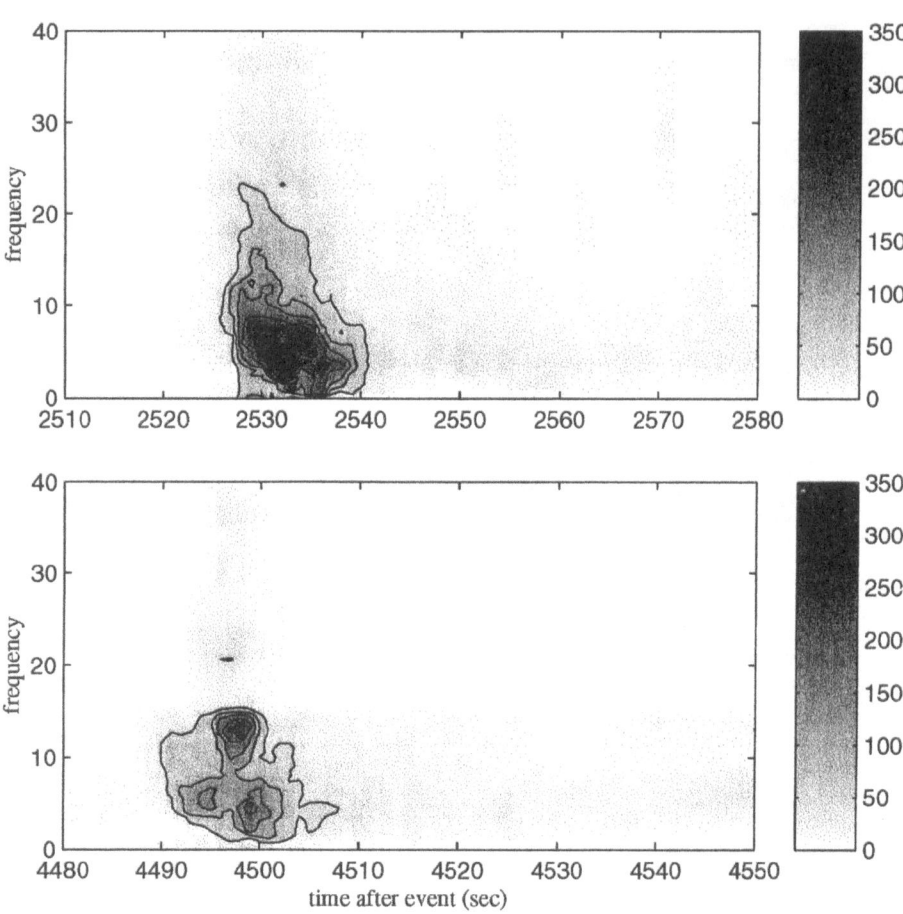

Figure 9
Spectral ratio sonograms for the Pt. Sur recordings of the Hawaii event, top, and the nuclear test, below,
formed by dividing sonograms for each time slice by average noise powers computed for quiet intervals
before the *T*-phase arrivals. Contours indicate intervals of 50.

T-phase amplitudes observed at the seismic stations are substantially lower for this
event than for the Hawaii event, with significant SNR only at SAO. This is only
partly attributable to the lower SNR of the acoustic *T* phase in the 2–4 Hz frequency
band, which couples most efficiently to the seismic *T* phase.

The considerably lower SNR of the nuclear tests, as compared to the Hawaii
events, is mainly attributable to poorer acoustic to seismic coupling at this azimuth.
Figure 10 shows the slope angle in the direction of acoustic propagation for this
event, along with geodesic paths from the source to each receiver. As indicated, at
this azimuth, most of the stations lie further from the steep sections of the continental
slope, thus the converted *T* phase must propagate further through the attenuative

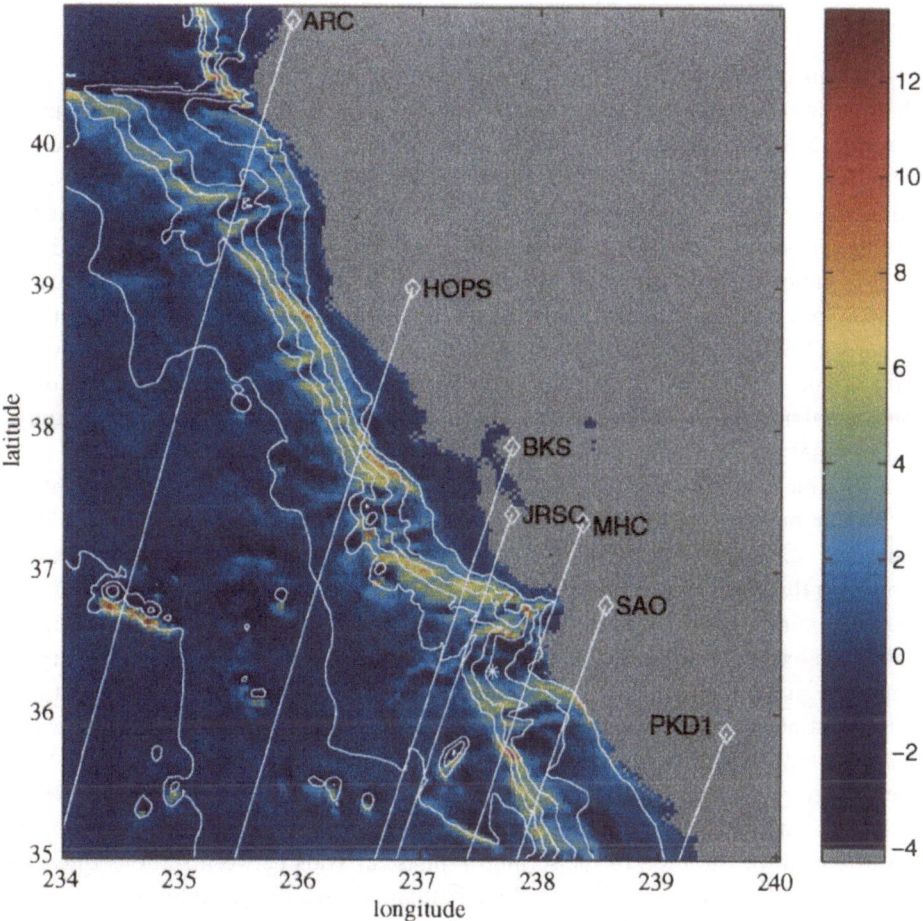

Figure 10
Slope angle in the direction of propagation from the Mururoa nuclear test to the receiver, shown to the same scale as in Figure 6. F = 2 Hz cutoff depths and geodesic paths from each station to the source are also shown.

crust. Since the acoustic phase strikes the continental slope at an oblique angle, the T phase undergoes lateral refraction as it couples from acoustic to seismic energy; the amount of deflection depends upon the velocity contrast, with P waves undergoing greater lateral refraction than S waves. Only at JRSC and SAO is the continental slope nearly perpendicular to the propagation direction, however, the slope is much shallower along this direction than for the Hawaii events. We therefore attribute the significantly lower SNR values of the T phases resulting from the nuclear tests to the extended distance between the coupling points and the receivers, and to the shallower rise of the continental slope along this azimuth.

Discussion and Conclusions

Current plans call for several T phase stations to be located on volcanic shield islands and for one at the edge of a continental margin (on Queen Charlotte Island, Canada). Seafloor slopes at volcanic islands are quite steep (12°–20°), and velocities are high, such that $v_s > v_w$ (e.g., PISERCHIA et al., 1998). At continental margins, the sediment-covered seafloor is characterized by very gentle inclinations (3°–6°) at the continental slope. Reflection and transmission energy flux densities computed for $v_s > v_w$ indicate that most acoustic energy will couple into S waves at volcanic islands, except at very steep slopes. This is consistent with the observations of HANSON (1998) at Ascension Island, of KOYANAGI et al. (1995) at Hawaii, and of TALANDIER and OKAL (1998) at Polynesian shield islands. At the edge of a continental margin, T waves convert to both P and S waves as observed in this study, and by SHAPIRA (1981), and CANSI and BETHOUX (1985).

A marine sediment-covered seafloor has similar acoustic properties to that of an equivalent fluid, i.e., one with identical density and compressional velocity. Thus coupling is expected to occur at distinct cutoff depths corresponding to modal cutoff depths. In this study, we do not discern distinct arrivals corresponding to each mode coupling point, and would not expect to, as the incident T phases last 15–20 seconds, which is greater than the travel time between mode coupling points. The largest amplitude arrivals occur at stations with propagation paths crossing steep portions of the continental slope near the station, where delay times between cutoff points are minimal and propagation through the crust is minimal. The largest arrivals thus likely correspond to the nearly simultaneous arrival of several converted modes so that modal information in the acoustic wavefield is lost in coupling to seismic energy. Variations in the duration of the seismic T-phase arrivals between stations thus can be attributed to variations in ocean floor bathymetry near each station.

Noise levels generally decrease with increasing distance from the coast so that seismic T phases can be detected extending several hundred kilometers inland. Comparison of T-phase SNRs over a 0.1–10 Hz band at the Berkeley array to those observed at Pt. Sur indicates that the seismic T-phase signal is lower at all frequencies than the acoustic T-phase arrival. Furthermore, signal strengths decrease more rapidly with increasing frequency, especially at stations located several hundreds of kilometers from the coast. The relatively high SNR at SAO, compared to the other seismic stations, may be attributed to the fact that the continental shelf is unusually narrow at this point, so that the low-mode coupling points lie closer to SAO than at the other sites. Furthermore, for the Hawaii events, and to a lesser extent for the nuclear tests, the conversion of acoustic to seismic energy along the direction of propagation to SAO occurs along a convex portion of the continental slope. Since the seismic phase is faster than the acoustic phase, this would focus the seismic phases near SAO. This may account for the high SNR at this station and slow decrease in SNR with increasing frequency, and suggests that 3-D effects may be important in accurately modelling acoustic to seismic conversions. In general, optimal sites for T-phase stations will be at locations where the continental slope lies close to the coastline. For slopes covered by marine sediments, acoustic to seismic coupling is efficient for slopes as low as 5°.

Acknowledgements

The seismic data were made available by the Berkeley Seismological Laboratory at the University of California, Berkeley through the Northern California Earthquake Data Center (NCEDC). The authors are grateful to Y. Gitterman and an anonymous reviewer for careful reviews. This work was supported by Defense Threat Reduction Agency grant number DSWA98-1-004.

References

BREKHOVSKIKH, L. M., and GODIN, O. A., *Acoustics of Layered Media I: Plane and Quasi-plane Waves* (Springer-Verlag, Berlin 1990).

BREKHOVSKIKH, L. M., and LYSANOV, Y. P., *Fundamentals of Ocean Acoustics* (Springer-Verlag, Berlin 1991).

CANSI, Y., and BETHOUX, N. (1985), *T Waves with Long Inland Paths: Synthetic Seismograms*, J. Geophys. Res. *90*, 5459–5465.

COLE, R. H., *Underwater Explosions* (Princeton Univ. Press, Princeton 1948).

COLLINS, M. D. (1993), *An Energy-conserving Parabolic Equation for Elastic Media*, J. Acoust. Soc. Am. *94*, 975–982.

HAMILTON, E. L. (1980), *Geocoustic Modeling of the Sea-floor*, J. Acoust. Soc. Am. *68*, 1313–1340.

HANSON, J. A., *Seismic and Hydroacoustic Investigations near Ascension Island* (Ph.D. Thesis, UCSD, San Diego 1998).

HEANEY, K. D., KUPERMAN, W. A., and MACDONALD, B. E. (1991), *Perth-Bermuda Sound Propagation (1960): Adiabatic Mode Interpretation*, J. Acoust. Soc. Am. *90*, 2586–2594.

JENSEN, F. B., and KUPERMAN, W. A. (1980), *Sound Propagation in a Wedge-shaped Ocean with a Penetrable Bottom*, J. Acoust. Soc. Am. *67*, 1564–1566.

JENSEN, F. B., and KUPERMAN, W. A., *Computational Ocean Acoustics* (American Institute of Physics Press, New York 1993).

KOYANAGI, S., AKI, K., BISWAS, N., and MAYEDA, K. (1995), *Inferred Attenuation from Site Effect-corrected T Phases Recorded on the Island of Hawaii*, Pure appl. geophys. *144*, 1–17.

MCMANUS, D. A., BURNS, R. E., VON DER BORCH, C., GOLL, R., MILOW, E. D., OLSSON, R. K., VALLIER, T., and WESER, O. (1970), *Site 32*, Init. Repts. D.S.D.P 5, 15–56.

MCMANUS, D. A., BURNS, R. E., VON DER BORCH, C., GOLL, R., MILOW, E. D., OLSSON, R. K., VALLIER, T., and WESER, O. (1970), *Site 33*, Init. Repts. D.S.D.P 5, 57–80.

NOLET, G., and DORMAN, L. M. (1996), *Waveform Analysis of Scholte Modes in Ocean Sediment Layers*, Geophys. J. Int. *125*, 385–396.

OFFICER, C. B., *Introduction to the Theory of Sound Transmission, with Application to the Ocean* (McGraw-Hill, New York 1958).

PISERCHIA, P.-F., VIRIEUX, J., RODRIGUES, D., GAFFET, S., and TALANDIER, J. (1998), *Hybrid Numerical Modelling of T-wave Propagation: Application to the Midplate Experiment*, Geophys. J. Int. *133*, 789–800.

SEIBOLD, E., and BERGER, W. H., *The Sea-floor* (Springer-Verlag, New York 1996).

SHAPIRA, A. (1981), *T Phases from Underwater Explosions off the Coast of Israel*, Bull. Seismol. Soc. Am. *71*, 1049–1059.

SMITH, W. H. F., and SANDWELL, D. T. (1997), *Global Sea-floor Topography from Satellite Altimetry and Ship Depth Soundings*, Science *277*, 1956–1962.

TALANDIER, J., and OKAL, E. A. (1998), *On the Mechanism of Conversion of Seismic Waves to and from T Waves in the Vicinity of Island Shores*, Bull. Seismol. Soc. Am. *88*, 621–632.

TOLSTOY, I., and EWING, M. (1950), *The T-phase of Shallow Focus Earthquakes*, Bull. Seismol. Soc. Am. *40*, 25–52.

YEATS, R. S., HAQ, B. U., BARRON, J. A., BUKRY, D., CROUCH, J. K., DENHAM, C., DOUGLAS, A. C., GRECHIN, V. I., LEINER, M., NEIM, A. R., PAL, S., PISCHIOTTO, K. A., POORE, R. Z., SHIBATA, T., and WOLFART, R. (1981), *Site 467: San Miguel Gap*, Init. Repts. D.S.D.P *63*, 23–50.

(Received June 30, 1999, revised November 10, 1999, accepted December 17, 1999)

To access this journal online:
http://www.birkhauser.ch

Pure appl. geophys. 158 (2001) 531–565
0033–4533/01/030531–35 $ 1.50 + 0.20/0

▌Pure and Applied Geophysics

Empirical and Numerical Modeling of T-phase Propagation from Ocean to Land

JEFFRY L. STEVENS,[1] G. ELI BAKER,[1] RON W. COOK,[1] GERALD L. D'SPAIN,[2]
LEWIS P. BERGER[2] and STEVEN M. DAY[3]

Abstract — T-phase propagation from ocean onto land is investigated by comparing data from hydrophones in the water column with data from the same events recorded on island and coastal seismometers. Several events located on Hawaii and the emerging seamount Loihi generated very large amplitude T phases that were recorded at both the preliminary IMS hydrophone station at Point Sur and land-based stations along the northern California coast. We use data from seismic stations operated by U. C. Berkeley along the coast of California, and from the PG&E coastal California seismic network, to estimate the T-phase transfer functions. The transfer function and predicted signal from the Loihi events are modeled with a composite technique, using normal mode-based numerical propagation codes to calculate the hydroacoustic pressure field and an elastic finite difference code to calculate the seismic propagation to land-based stations. The modal code is used to calculate the acoustic pressure and particle velocity fields in the ocean off the California coast, which is used as input to the finite difference code TRES to model propagation onto land. We find both empirically and in the calculations that T phases observed near the conversion point consist primarily of surface waves, although the T phases propagate as P waves after the surface waves attenuate. Surface wave conversion occurs farther offshore and over a longer region than body wave conversion, which has the effect that surface waves may arrive at coastal stations before body waves. We also look at the nature of T phases after conversion from ocean to land by examining far inland T phases. We find that T phases propagate primarily as P waves once they are well inland from the coast, and can be observed in some cases hundreds of kilometers inland. T-phase conversion attenuates higher frequencies, however we find that high frequency energy from underwater explosion sources can still be observed at T-phase stations.

Key words: Hydroacoustic, T phase, CTBT.

Introduction

The International Monitoring System (IMS) hydroacoustic network will consist of the 6 underwater hydroacoustic stations and 5 land-based seismic T-phase stations shown in Figure 1. The hydroacoustic stations are decidedly more sensitive to

---
[1] Maxwell Technologies, Systems Division, 9210 Sky Park Court, San Diego, CA 92123, USA. E-mail: Stevens@maxwell.com
[2] Marine Physical Laboratory, Scripps Institution of Oceanography, San Diego, CA, USA.
[3] San Diego State University, San Diego, CA, USA.

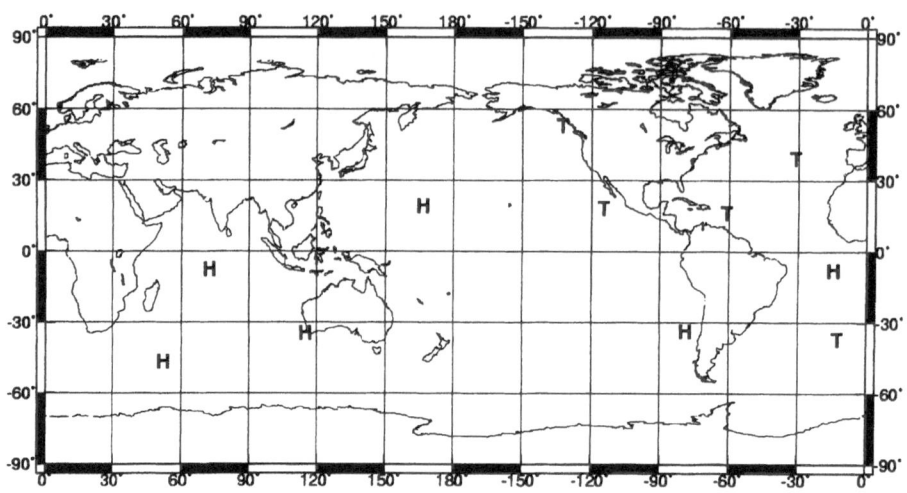

Figure 1
Hydroacoustic (*H*) and *T*-phase (*T*) stations in the future IMS network.

underwater signals than the *T*-phase stations and have a higher sampling rate and broader frequency range. The broader frequency range is important for identifying explosions, which are characterized by higher frequency content than other sources. Because of these limitations, it is important to understand the efficiency of *T*-phase conversion in order to assess the capabilities of the IMS network for detection and identification of underwater sources.

Currently, only part of the hydroacoustic network is in place, and an additional hydroacoustic station at Point Sur acts as part of the IMS network. Point Sur will not be part of the IMS once the final network is in place. Figure 2 shows the location of the two hydroacoustic and one *T*-phase stations which are currently operating in and adjacent to the Pacific Ocean. Also shown on the map are the locations of the emerging seamount Loihi (UNIVERSITY OF HAWAII, 1998), which has been the source of strong *T* phases, and the location of the source of *H* phases (explosion-like *T* phases) that occurred in November, 1997. Data from these two sets of events are discussed in the following sections of this paper.

In this paper, we investigate the nature of the hydroacoustic to *T*-phase conversion process in the following ways:

1. We compare data from hydrophones in the water column with data from the same events recorded on island and coastal seismometers, using data from the Loihi events. This gives us a direct measurement of the *T*-phase transfer function.

2. We perform two-dimensional finite difference calculations of the propagation of the hydroacoustic signal onto the coast. These are then compared with the observed signals and used to determine where conversion occurs and what types of seismic waves are generated by the conversions.

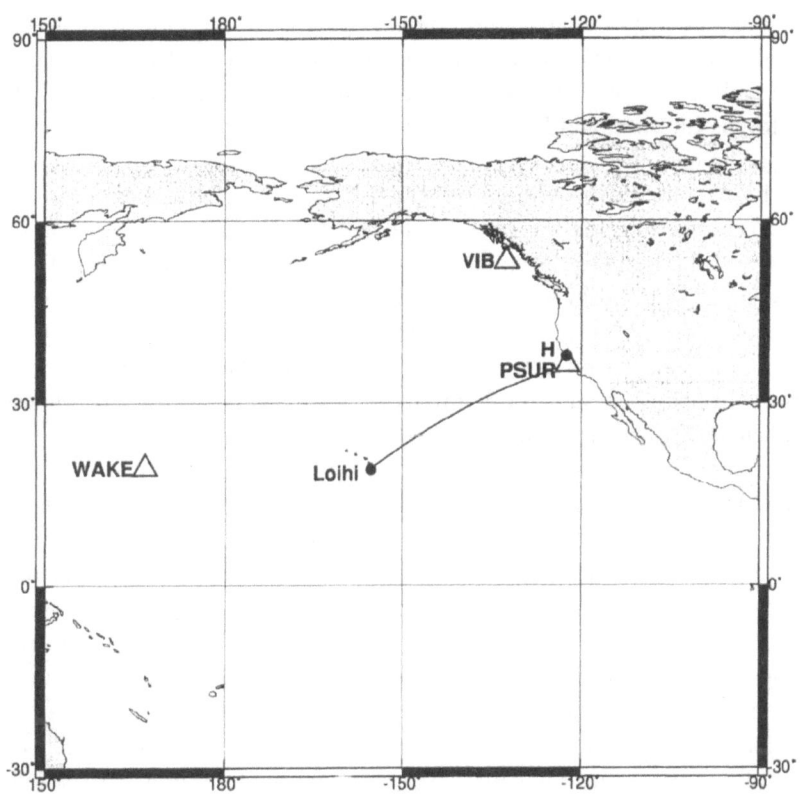

Figure 2
Hydroacoustic network stations in the current International Monitoring System in the Pacific Ocean. VIB is a *T*-phase station. Point Sur and Wake are underwater hydroacoustic stations. "Loihi" marks the location of strong *T*-phase sources south of Hawaii used in this study. "H" marks the location of a series of underwater explosions. The great circle path from Loihi to Point Sur is also shown.

3. We look at signals from small underwater explosions recorded on two underwater hydrophone stations and the IMS *T*-phase station at VIB to determine whether the high frequency energy characteristic of explosions is observable after *T*-phase conversion.

4. We examine propagation of *T* phases in California from coastal stations to stations located up to 200 km inland, and survey *T* phases that have propagated much farther inland in Australia and South America to determine the nature of the *T* phase as it propagates on land.

Transmission of T-phase Energy from the Ocean onto Land

We have gathered data sets from events that were recorded on both underwater and coastal seismic stations for the purpose of directly measuring *T*-phase conversion. Several earthquakes from the emerging seamount Loihi and on the island of Hawaii generated very strong *T* phases that impacted the California coast. The location and size of these events, taken from the Reviewed Event Bulletin (REB) of the Prototype International Data Center, are listed in Table 1, and the locations are shown on the map in Figure 3.

We have collected recordings of these events from the Point Sur (PSUR) hydroacoustic station, the Pacific Gas and Electric Central Coast seismic network (PG&E), and the Berkeley broadband digital seismic network (BDSN). The PG&E network is used primarily to monitor seismicity near the Diablo Canyon nuclear power plant. The location of these stations and the bathymetry of the California continental shelf are shown in Figure 4. This is a nearly optimal situation for study of *T*-phase conversion because the hydroacoustic waves impact the coastline almost perpendicular to the coast as indicated by the ray path on Figure 4.

The sampling rate and instrumentation are different for each of the three networks. Point Sur is sampled at 200 samples per second. A routine available from the PIDC converts the data to pressure in micropascals over a frequency band of approximately 3 Hz to 80 Hz. The routine also acts as a band-pass filter, removing frequencies outside of this frequency band. From this, we determine the instrument response, which to a very good approximation is linearly proportional to frequency at frequencies less than 25 Hz. Data in the BDSN network is sampled at 20 samples per second. BDSN instrument responses were provided with the data, and are approximately flat to velocity in the 1–10 Hz frequency band. The PG&E network consists of S-13 seismometers with a sampling rate of 100 samples per second. The instrument at the PG&E stations is flat to velocity from 2 to 25 Hz, with a low-pass filter above 25 Hz. We have gain corrections for each of the PG&E instruments, however we do not know the resulting units after the gain corrections are applied. We have made the assumption that the gains correct to 0.01 mm/sec since that makes the amplitudes consistent with the BDSN amplitudes, however any conclusions drawn from these measurements must take into account that the absolute amplitudes are uncertain. The IMS *T*-phase stations such as VIB also have S-13 seismometers and a sampling rate of 100 samples per second, however they do not have the low-pass filter at 25 Hz. Owing to the differences in instrumentation, the usable frequency band for obtaining *T*-phase transfer functions from this complete data set is about 2–9 Hz.

Table 1

Origins of Loihi/Hawaii events used in this study

Event #	Origin ID	Date	Time (GMT)	Latitude	Longitude	m_b
1	751298	1996/07/23	13:24:59	18.9620	−155.3959	4.40
2	752002	1996/07/24	17:38:50	18.9833	−155.4213	4.51
3	1074833	1997/06/30	15:47:38	19.3108	−155.1058	5.06
4	1110057	1997/08/15	01:54:38	19.4083	−155.1098	4.43

Four Quakes near Hawaii

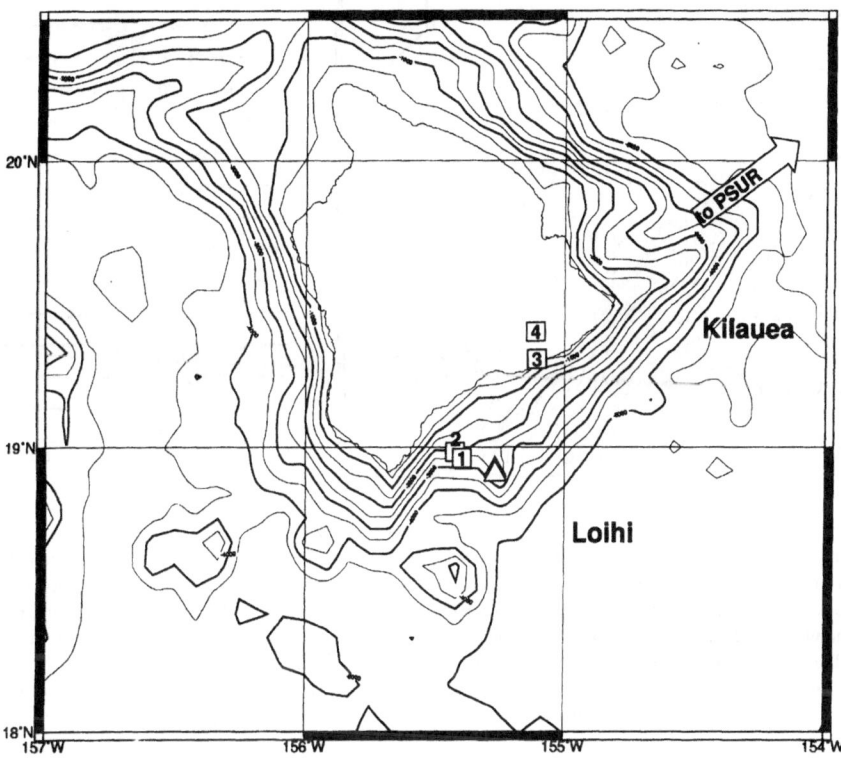

Figure 3

Map of the Hawaiian Islands and surrounding ocean bottom bathymetry. The squares labeled 1 through 4 (the square for event 2 is covered by that for event 1) denote the epicentral locations of Hawaiian events 1 through 4 given in Table 1 as obtained from the Reviewed Event Bulletin. The two triangles (which are co-located on this spatial scale) show the corresponding epicentral locations for events 1 and 2 in 1996 derived from Hawaii Volcano Observatory local station recordings (CAPLAN-AUERBACH, 1999). The bathymetry information, obtained from the GMT database (NOAA, 1998), are contoured in 500 meter increments.

Figure 5 presents a sample of the data recorded at one station in each network for event #1. The data have been band-pass filtered from 3–8 Hz. There are several interesting features in these data. The Point Sur and PG&E data are very similar in shape and duration. The Berkeley data however are more spread out and dispersed, with large amplitudes lasting nearly three times as long as the signals at the other two stations. The Point Sur record also exhibits a fairly large second arrival, which is not apparent on the other records. This may be a reflected arrival from the California Coast.

We have calculated *T*-phase transfer functions for this data set by taking the ratio of the instrument corrected spectral amplitudes of the seismic stations to the Point Sur station for the events listed in Table 1 for all good quality data at the PG&E

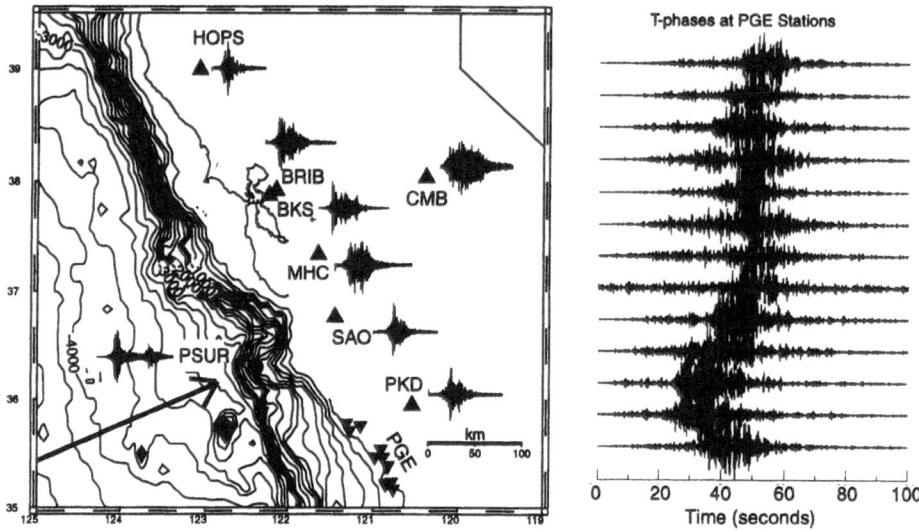

Figure 4

Records from the June 30, 1997 Loihi event and station locations. The propagation direction is indicated by
the arrow, which points to the hydroacoustic station PSUR (indicated by the circle). 150 seconds of the *T*-
phase record is shown to the station's left. The BDSN stations (large triangles) and corresponding records
are included on the map on the same scale. The PG&E station locations are indicated by inverted triangles
and their waveforms are shown to the right, in order of increasing distance from bottom to top. The arrival
times do not correspond well to total distance because the distance from the conversion point varies. The
bathymetry is contoured in 200 meter intervals.

stations and the 6 coastal BDSN stations (CMB was excluded because it is much
farther inland). The spectral ratios were averaged for all events for each station. Two
stations recorded all four events, 7 stations recorded 3, 9 stations recorded 2, and two
stations recorded only 1 event. The resulting spectral ratios are shown in Figure 6.
There is a clear decline in amplitude with frequency for both networks, but it is
particularly pronounced for the BDSN network. This is likely due to greater
attenuation of the higher frequencies because of the longer paths to the BDSN
stations. Simulations discussed later in this paper indicate that considerable energy in
the hydroacoustic wave is transmitted to land at approximately 200 m bathymetry,
which corresponds to the contour line (Fig. 4) nearest the coast. The crustal paths
from the conversion point to the PG&E stations are only 5–20 km long, while paths
to the BDSN stations are 40–75 km.

Although the transfer function can only be calculated by numerical methods, we
estimate an approximate upper bound on the transfer function by considering the
simple case of a plane acoustic wave travelling in the ocean and propagating into a
solid at normal incidence. This is an approximate upper bound in the sense that any
scattering or other boundary effects will reduce the amplitude below this level,
although conversion to shear waves or focusing could cause amplitudes to exceed this

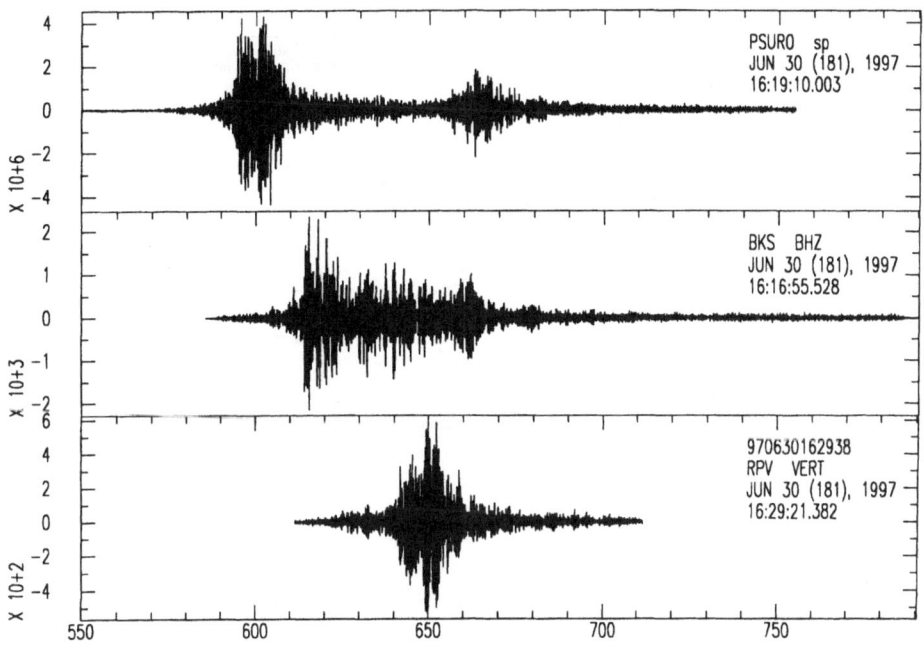

Figure 5

Data for the June 30, 1997 event recorded at Point Sur (Top), Berkeley (Center), and RPV in the PG&E network (Bottom). The horizontal axis is time in seconds and the start time is the same for all three seismograms.

value. For a plane acoustic wave, the pressure is related to the velocity at a point by $P = \rho \alpha V$ where α is compressional velocity, ρ is density and V is particle velocity in the direction of propagation. For propagation of a plane wave at normal incidence from a fluid into a solid, the reflection coefficient R and transmission coefficient T for velocity are given by (e.g., ACHENBACH, 1973):

$$R = \frac{\rho_s \alpha_s - \rho_f \alpha_f}{\rho_s \alpha_s + \rho_f \alpha_f} \tag{1}$$

$$T = \frac{2 \rho_f \alpha_f}{\rho_s \alpha_s + \rho_f \alpha_f} \tag{2}$$

where the subscript s indicates solid and f indicates fluid. The transmission coefficient for propagation of pressure in the fluid to velocity in the solid is therefore:

$$T_{P \rightarrow V} = \frac{2}{\rho_s \alpha_s + \rho_f \alpha_f} \ . \tag{3}$$

Table 2 lists a velocity model for the California coast near Point Sur from MOONEY *et al.* (1998) with the shear velocity as modified by STEVENS and McLAUGHLIN (1997). Using a typical water velocity and density of 1480 m/s and

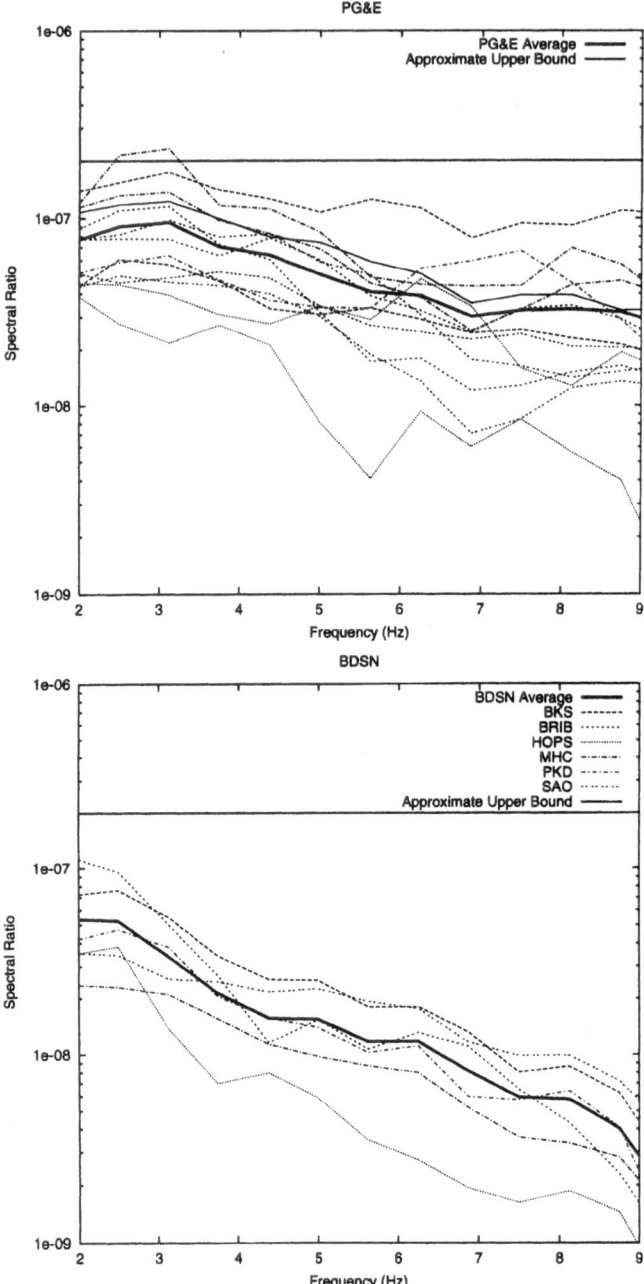

Table 2

Velocity model for the California coast

Depth (m)	P velocity (m/s)	S velocity (m/s)	Density (kg/m^3)
500	3400	1500	2200
1500	5000	2900	2500
11000	6100	3484	2750
21000	6300	3503	2800
31000	6600	3464	2900
∞	8000	4260	3300

1000 kg/m^3, respectively, we can estimate upper bounds on the velocity transfer function of 0.21 and 0.36 for the top two layers of the coastal California earth model. The pressure to velocity transfer function for each layer has an upper bound of 1.4×10^{-7} and 2.4×10^{-7} meters/second/Pascal. The heavy lines in Figure 6 delineate the average transfer function for all of the coastal stations in the BDSN and PG&E networks. At 2 Hz, the BDSN spectra are a factor of 3–4 less than the estimate given above, which gives a numerical estimate of the efficiency of *T*-phase transmission into the coast. That is, the BDSN records indicate that at low frequencies the transfer function is reduced by complex coastal interactions by a maximum of a factor of 3–4. The spectra of the BDSN records fall off by an order of magnitude over the 2–9 Hz frequency band. The PG&E records show that for stations close to the coast, the maximum attenuation with frequency is a factor of 3–4 over the 2–9 Hz frequency band.

Far Inland T Phases

The ocean hydroacoustic phase may convert to a variety of complex phases due to the interaction with the coast. We can gain insight into the type of conversion that occurs by looking at far inland *T* phases, and examining the propagation speeds of these waves. The *T* phase from the 1997/06/30 event is large enough to be seen well inland from the California coast. In Figure 7, we show the relative amplitudes of

◄

Figure 6

Ratios of seismic spectra at stations in California that recorded the Loihi events, divided by the Point Sur spectra. These are T-phase transfer functions converting pressure to vertical velocity. The frequency band shown here is from 2 to 9 Hz. The top figure shows PG&E stations and the bottom figure shows BDSN stations. The absolute amplitude of the PG&E records is uncertain, however the amplitude of the BDSN records, and the spectral shape of both sets of curves, are accurate. Also shown is the approximate upper bound for conversion from pressure in the ocean to particle velocity on land as discussed in the text. The heavy line is the average log spectral ratio for all stations except far inland stations. Spectral ratios are in MKS units (meters/second/Pascal).

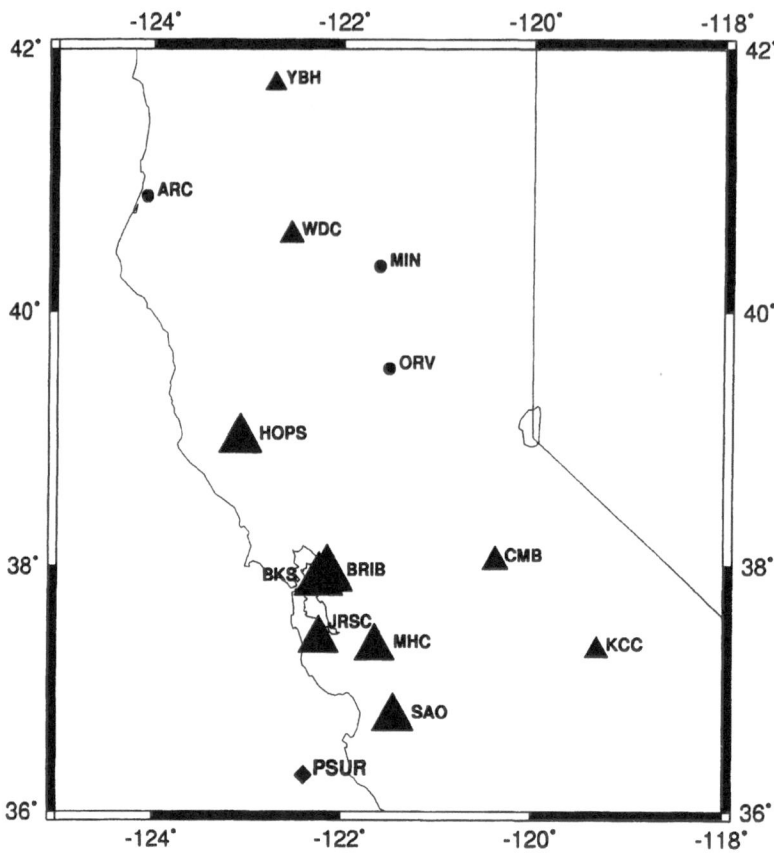

Figure 7

T-phase amplitudes at stations in the BDSN network. Symbol sizes are proportional to log amplitude. T phases were not measurable at stations indicated by circles.

T-phase arrivals from this event at the BDSN stations. T phases were observed as much as 200 km inland from the coast.

Figure 8 shows two sets of seismograms recorded along the two (approximate) great circle paths (see Fig. 7). The paths are from SAO to KCC and from JRSC to MHC to CMB. Since the waveform is dispersed, there is uncertainty about when to pick the arrival, however if we use the peaks of each wavetrain, then we derive a velocity between SAO and KCC of 6.8 km/sec, and a velocity along the path of the other three stations of 5.6 km/sec. These velocities correspond to P-wave speeds, therefore the T phases must be travelling as P waves over this range.

Particle motion provides an independent means of assessing wave type. Figure 9 shows that the particle motion for the largest part of the record, near the front of the T phase, is strongly linear. This part of the record is clearly dominated by body waves, most likely P waves, in agreement with the conclusion from travel times.

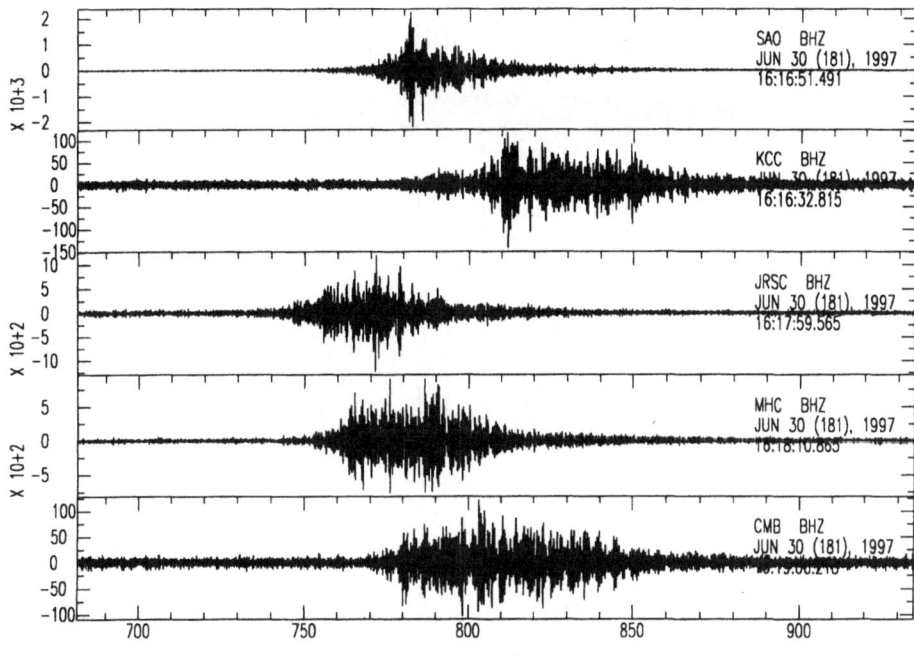

Figure 8

T phases recorded near the coast and far inland from the 1997 June 30 Loihi event. The top two seismograms and the bottom three seismograms were recorded at stations along approximately the same great circle path. The horizontal axis is time in seconds. The data have been high-pass filtered at 2 Hz.

Linearity decreases later in the record, although it is not clear whether that is due to the arrival of later scattered *P*-wave energy, surface wave arrivals, or a mixture of phases. The particle motion of the earliest part of the *T* phase is elliptical and retrograde, suggesting that this part of the wavetrain is composed of Rayleigh waves. This is consistent with the simulations later in this paper, which indicate that surface waves can precede the *P* wave near the coast, due to earlier conversion of the *T* phase in water to surface waves.

COOK and STEVENS (1998) collected a number of examples of *T* phases recorded far inland from the coast. We present two examples of these here. In Figure 10, we show the locations of stations that recorded *T* phases from an earthquake south of Australia, and the bathymetry and topography along the path. The path on land to the most distant station, ASPA, exceeds 1000 km. The travel times for all records are consistent with propagation on land at P_n velocity. The velocity labeled "AV" on the plot is the apparent velocity in the water after subtracting the travel time for propagation on land. This apparent velocity acts as a consistency check and should be close to the velocity of water if the *T* phase is traveling on land at P_n speeds. This is additional evidence that *T*-phase propagation on land, at least once the wave has propagated well inland from the coast, consists of *P* waves.

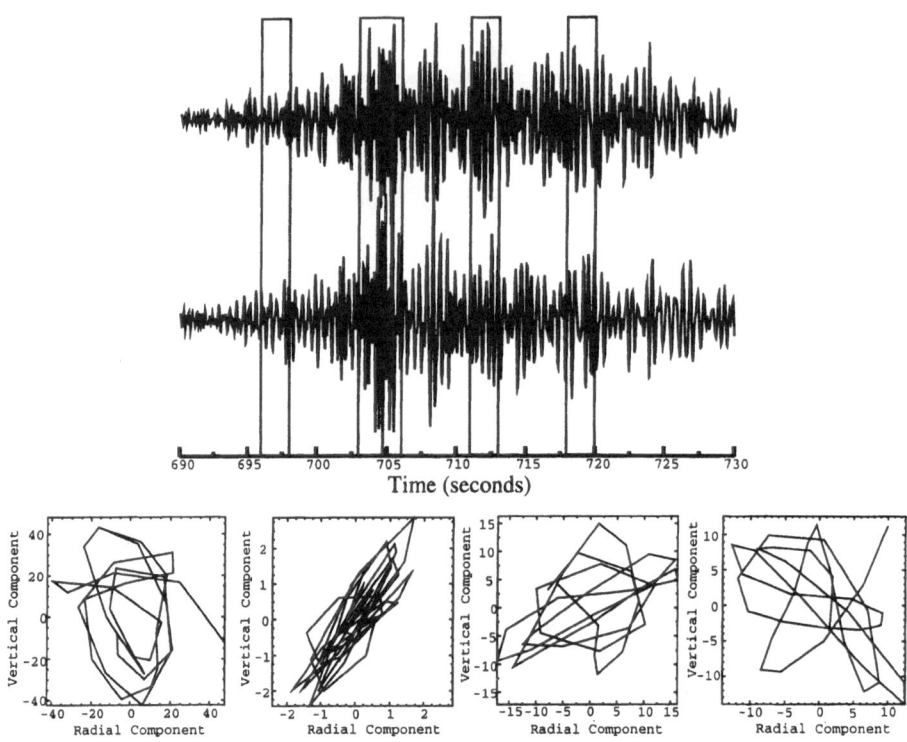

Figure 9

Vertical (top trace) and radial (lower trace) component seismograms from SAO for the 1997 June 30 event. Particle motions for four time windows outlined by shading are shown below the seismograms. The very linear motion of the second time window indicates dominance there by body waves, and the elliptical motion of the first time window suggests that the early part of the wavetrain is composed of surface waves.

Figure 11 shows a second example, this time of a T phase from an earthquake in the eastern South Pacific Ocean recorded in Argentina, on a path that traveled through the Andes and over a distance of approximately 300 km on land. Far inland T phases have very peculiar travel times because they travel very slowly (~1.4 km/s) in the ocean, and very fast (up to 8 km/sec) as P waves on land. Consequently, the T phase starts out far behind the other seismic phases, but will eventually catch up to the Rayleigh wave if it travels far enough. Because of this, there is danger in a semi-automated processing system like the IDC that T phases could be misidentified as other phases. One approach to solving this problem would be to calculate the T-phase travel time for each event and if it is found, to associate the T phase with the correct event.

IMS Recordings of Small Explosion Sources

On November 10, 1997, 20 "H phases" were reported in the Reviewed Event Bulletin. "H phases" refer to T phases which have the characteristics of explosions —

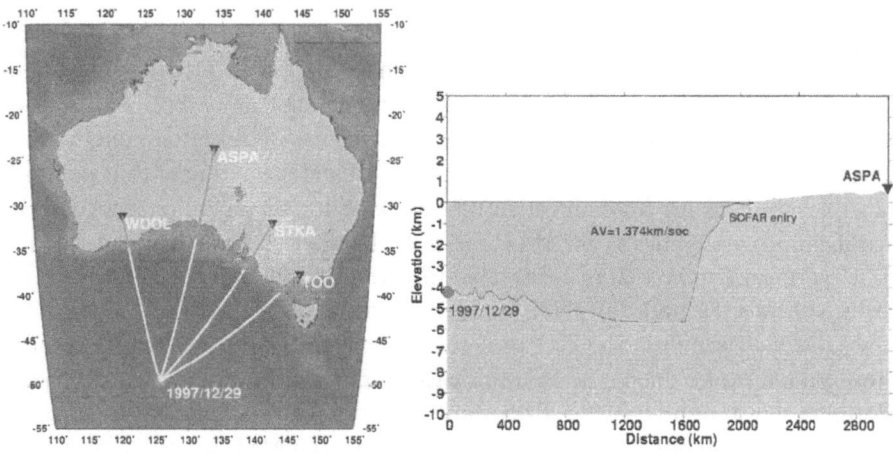

Figure 10

T phases were recorded far inland for the 1997/12/29 earthquake south of Australia. The left-hand figure shows a map of the source to receiver paths, and the right-hand figure shows bathymetry and topography along the path. The circle next to the date is the epicentral location.

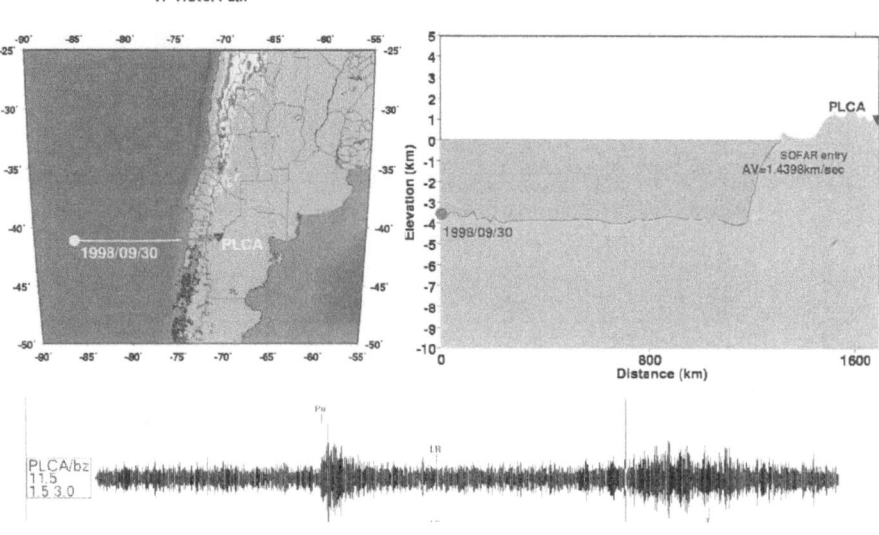

Figure 11

Map (left) and bathymetry/topography (right) for the path of *T* phases from the earthquake of 1998/09/30 recorded at PLCA. At the bottom is the seismogram recorded at PLCA showing with the phases marked.

short duration and enhanced high frequencies. These arrivals came from events that
were located off the coast of San Francisco as shown on the map in Figure 12.

These *H* phases were observed not only at Point Sur but also at Wake Island,
nearly 10,000 km away, and at VIB, the *T*-phase station in Canada (see Fig. 2).
These arrivals were subsequently identified as originating from a series of explosions
detonated off the coast of California as part of a submarine detection exercise. There
was a total of 60 explosions detonated, all identical and consisting of four pounds of
C4 explosive. This provides an interesting test case for the detection, location, and
identification capabilities of the IMS. First, the IMS and the processing done at the
PIDC performed very well in being able to detect and identify as explosions these
very small tests. Second, the ability to see and identify these events at Wake Island
shows just how well these signals travel in the ocean on unobstructed paths. Third, if
a four pound explosion can be seen this easily, an explosion in the ocean at the one
kiloton design threshold of the IMS should be impossible to miss unless there are
severe path obstructions (OLIVER and EWING, 1958, reported that the *T* phase from
the 30 kiloton underwater explosion Wigwam was large enough to be felt in
California and Hawaii). However, as is apparent from Figure 12, locating small
events is a more difficult problem. The PIDC event locations were distributed over a

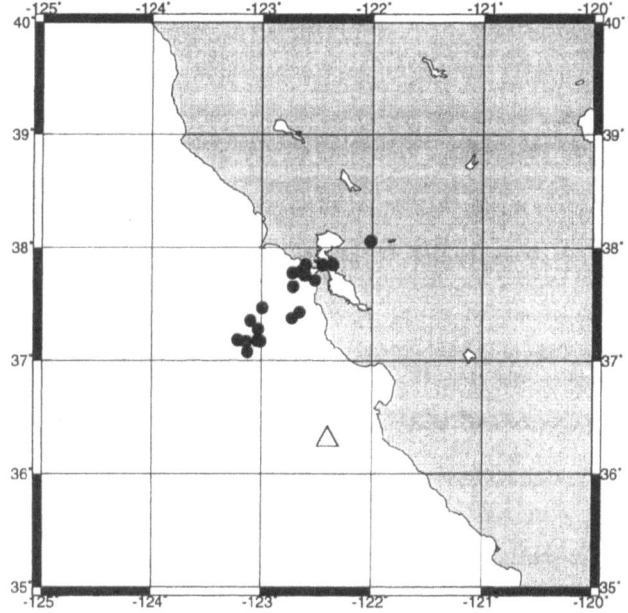

Figure 12
Locations of 1997/11/10 events generating *H* phases reported in the REB.

wide arc, with some events actually located on land. Although we do not know the exact location of the explosions, there are clearly large location errors.

Figure 13 shows arrivals from two of the explosions recorded at the three IMS stations. The character of the waveforms is distinctly different at each station. Wake, in particular, is especially more emergent, and has a maximum at the end of the wavetrain. These events also provide an interesting test of whether the high frequencies characteristic of explosion sources can be observed at *T*-phase stations. Figure 14 shows spectrograms of two waveforms, one recorded at Point Sur and one at VIB. The spectrogram at Point Sur is exactly what is expected from an explosion: short duration and strong high frequency energy, extending across the entire pass band of the instrument to near 90 Hz. The VIB spectrogram also registers strong high frequency energy extending across the pass band of the instrument to about 40 Hz. This shows that, for this example and similar cases, waveforms from explosions in the water will contain enough high frequency energy after propagation to *T*-phase stations to identify them as explosions.

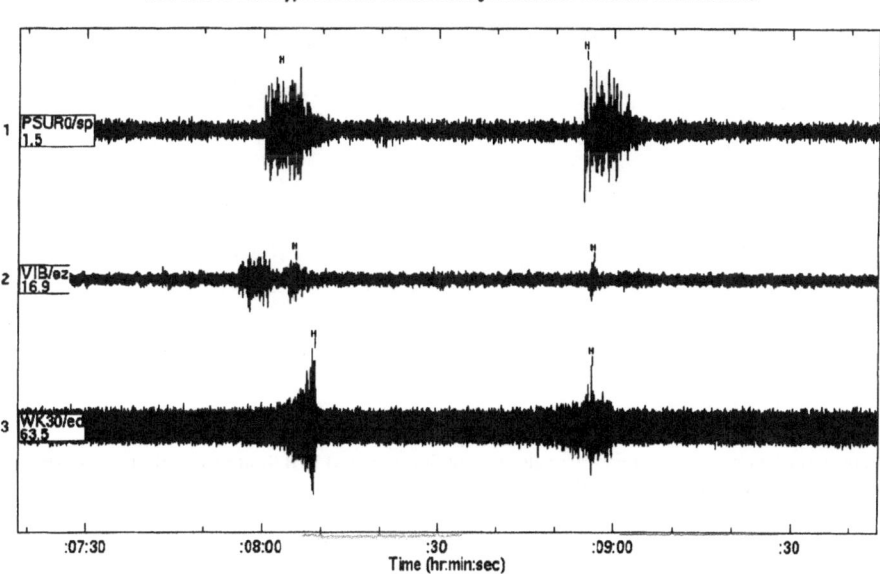

Figure 13

H-phase arrivals associated with the two events at Point Sur (top), VIB (middle) and Wake Island (bottom). "*H*" marks the arrival times listed in the REB. The time shown is the time after 1997/11/10 10:20 at PSUR. VIB and WAKE are advanced by the predicted travel time relative to PSUR. The absolute times of the first marked *H* in each trace are 10:28:03 (PSUR), 10:47:16 (VIB), and 11:45:35 (WAKE). Although the arrivals at VIB are less distinct than the other stations, the consistency with the arrival times of the signals from multiple events at the other two stations makes it unlikely that they have been misidentified.

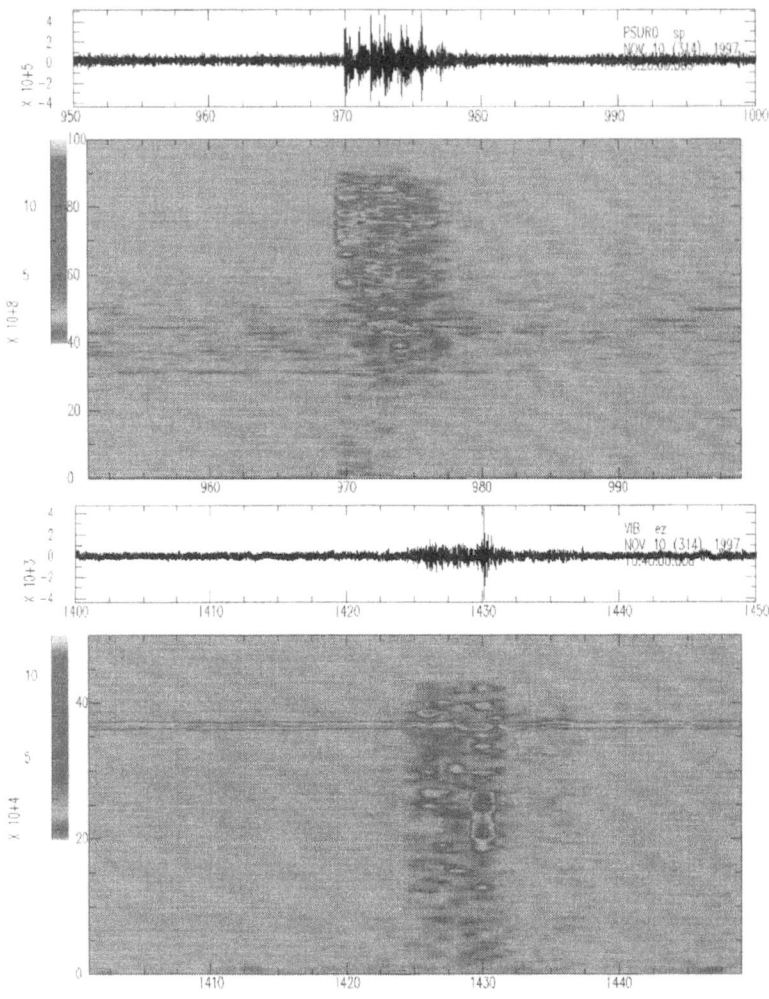

Figure 14

Spectrograms of one of the *H*-phase arrivals at Point Sur (top) and VIB (bottom). Both stations exhibit the strong high frequencies characteristic of underwater explosions. The horizontal axis is time in seconds.

Numerical Simulations of T-phase Transfer Functions

In order to better understand the *T*-phase transfer functions, and the nature of the seismic *T* phase as compared to the direct hydroacoustic signal, we have performed numerical simulations of the waveforms and the transition process. Because of the excellent data set for the Loihi events, we are modeling waveforms that travel from Loihi to the California coast traversing the location of the Point Sur hydroacoustic station. Our approach has been to first use these data to create empirical transfer functions relating the in-water acoustic pressure spectrum to the land vertical velocity component spectrum (Fig. 6), and subsequently to develop a numerical modeling capability to 1) identify those features in the arrival structure at the underwater stations that are associated with deep water propagation, and 2) determine the important geoacoustic parameters dictating the efficiency of water-to-land coupling. An early version of this work is described in STEVENS *et al.* (1998).

The map in Figure 2 indicates the locations of the Loihi seamount southeast of the Hawaiian Island chain and the Pt. Sur station on the continental slope just to the southwest of Monterey Bay on the central California coast. The two locations are separated by a distance of 3745 km. Bottom bathymetry data along the path was obtained from the National Geophysical Data Center (NOAA, 1998). The water depth is nominally 5000 m along the propagation path until the continental slope along the west coast of California is reached. The water depth at the Pt. Sur station is slightly less than 1400 m. Fifteen archived conductivity-temperature-depth (CTD) casts (SCRIPPS, 1998) at locations along the path between Loihi and Point Sur are used to obtain the 15 ocean sound speed profiles presented in Figure 15. The profiles illustrate the typical character of the deep sound channel ("SOFAR" channel) in Northeast Pacific environments, i.e., a sound speed minimum of 1480–1487 m/sec at a depth of 500 to 800 m. Differences in the profiles, most predominant at depths shallower than 1 km, reflect both seasonal changes as well as spatial variations along the propagation path. In the initial modeling efforts, attention has been focused on the effects of the shoaling bottom bathymetry at the continental slope. Therefore, in the region being modeled near Point Sur, a single typical profile with a sound speed of 1525 m/sec at the surface and 1482 m/sec at the sound channel axis at 700 m depth was used. Nine distinct temperature profiles were used to model the path between Loihi and Point Sur. A correction to account for earth curvature (BISWAS and KNOPOFF, 1970) is applied to the sound speed profile before it is used in the propagation codes.

Kraken, a normal mode code written by M. Porter, was used to calculate the hydroacoustic wave field off the coast of California. Kraken can be obtained through the Ocean Acoustics Library web site at http://oalib.njit.edu/. To model the range dependence, the propagation path has been divided into 48 consecutive range-independent segments. The range intervals for the segments are determined by the places where the bottom bathymetry contours change by 100 m. Eighteen segments were used on the first 3710 km of the path, to the point where the solution was transferred to the finite difference calculation. The ocean bottom for the in-water propagation calculations is modeled as a lossy fluid layer of 1-km thickness with a compressional velocity of 2100 m/sec and density of 2100 kg/m^3 overlying a lossy fluid halfspace with velocity 6000 m/sec and density 2700 kg/m^3.

The *T*-phase source is modeled as a single omnidirectional point at a depth of 1 km, corresponding to the depth of the peak of the Loihi seamount. Finite source dimensions and other near-source effects, including those such as scattering are

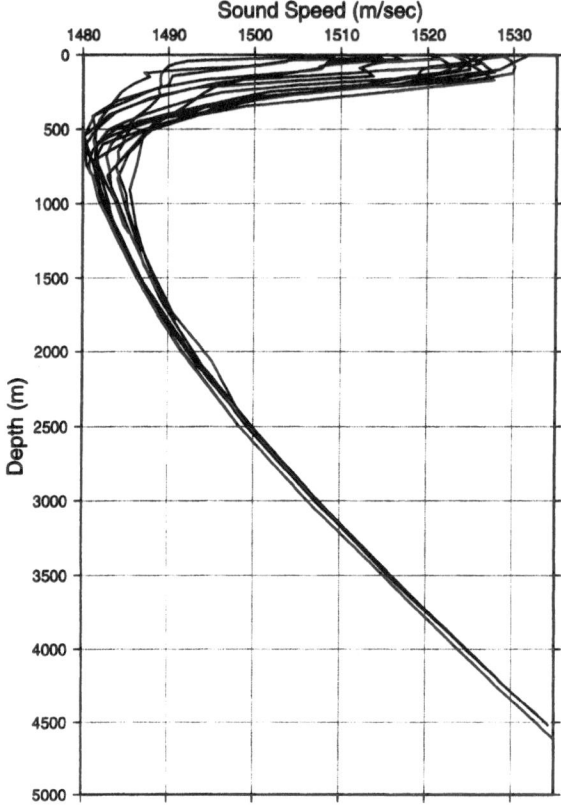

Figure 15
Fifteen sound speed profiles along the Loihi-to-Pt. Sur great circle path, derived from historical CTD casts.

neglected in this simulation. The close proximity of the Loihi seamount peak to the depth of the deep sound channel axis may provide a direct coupling of seismic energy into the deep sound channel, minimizing the importance of the field interactions in the region surrounding the source.

The calculations were done for single frequencies, at equally-spaced frequency increments from 0.4 Hz to 10 Hz. Figure 16 shows the acoustic pressure amplitude for the first 5 modes at 3 Hz, as a function of depth at a distance of 3710 km.

The adiabatic approximation (e.g., JENSEN *et al.*, 1993) has been used in the normal mode numerical modeling to model propagation through the deep ocean. The problem is transferred to a finite difference code at the point where the ocean depth begins to decrease rapidly. The ocean bottom in this part of the calculation is shown in Figure 17, together with the number of modes that can exist over this

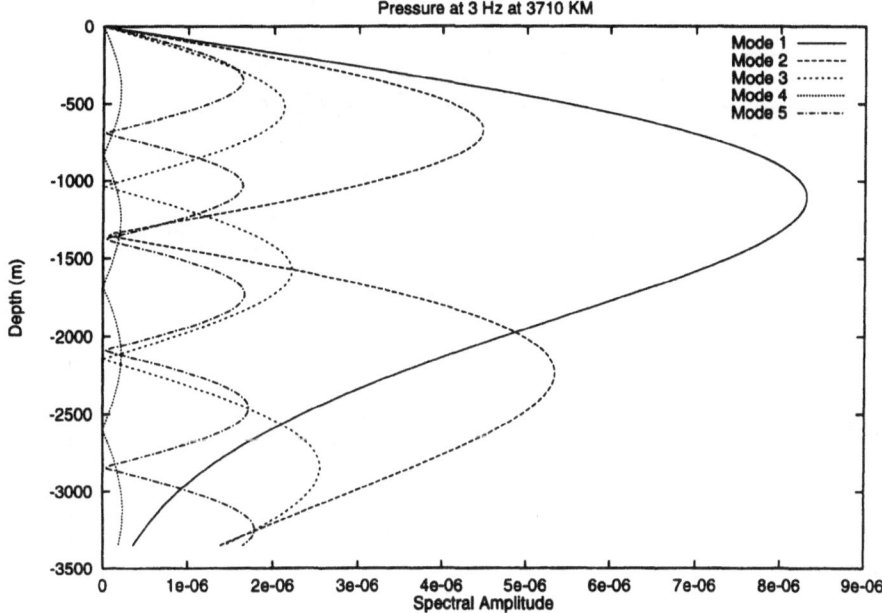

Figure 16
Spectral amplitude of the first five modes at a frequency of 3 Hz shown as a function of depth.

range. As the ocean waveguide becomes progressively shallower, the water depth where a given mode at a given frequency reaches cutoff determines the location along the continental slope where that mode's energy couples into the land seismic field. Figure 17 shows the number of modes that can propagate in the water column at the ten frequencies equally spaced from 1 to 10 Hz as a function of range. Most of the modes couple into the bottom over the 80-km range from 3710 to 3790 km, with the majority of the higher order modes coupling in over the 35-km interval from 3715 to 3750 km, and the lower order modes in the 20-km interval from 3765 to 3785 km. It is over the corresponding depth interval for this latter range, i.e., depths shallower than 1000 m, that the elastic properties of the continental slope are most critical in modeling the ocean-acoustic-to-land-seismic field coupling. A more detailed discussion of modal propagation in the ocean and modal coupling is given in D'SPAIN *et al.* (1999).

Two-dimensional Finite Difference Calculation of Propagation onto Land

We use the two-dimensional finite difference code TRES2D to model the propagation of hydroacoustic waves onto the coast. TRES is a finite difference code developed at Maxwell Technologies (S-CUBED) with 2-D and 3-D versions that are

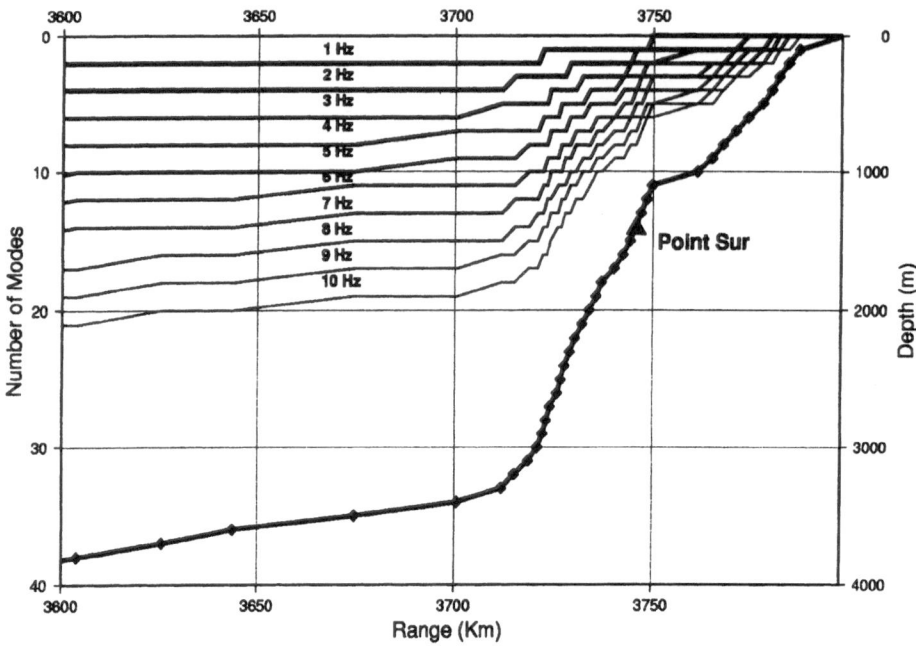

Figure 17
Number of in-water propagating modes at each frequency versus range. The bottom bathymetry also is
shown as a set of connected diamonds. Each curve corresponds to a different frequency. The top curve
corresponds to a frequency of one Hz, the second to two Hz, etc. At 10 Hz, 21 modes are supported at the
deepest depth, while only two modes exist at 1 Hz.

used to model wave propagation (MCLAUGHLIN and DAY, 1994). In this case we
model a 2-D slice of the propagation path. A series of 2-D slices can be used to
approximate the 3-D seismic wave field as long as off-axis arrivals are not important.
A full 3-D calculation would be necessary to model off-axis scattering. The Kraken
normal mode solution in the ocean at a distance of 3710 km from Loihi was used as
the source in the finite difference calculation using the representation theorem as
described in Appendix A. The first step in this process is to convert the modal spectra
to time domain displacements and pressures at each node point (we actually calculate
the derivative of these quantities, so that the output of the finite difference calculation
is velocity). The phase and group velocities of the first five modes are shown in
Figure 18. The first arrival of each mode, which also has the maximum amplitude,
travels at the maximum velocity as indicated by the peak of the group velocity curve.
The long, low frequency tail of each dispersion curve presents a problem for the
calculation, because at a distance of 3710 km these cause wavetrains that rattle
on considerably longer than the duration of the calculation. This part of the
dispersion curve corresponds to lower frequencies that feel the ocean bottom, and
consequently are more heavily attenuated than frequencies that are confined to the

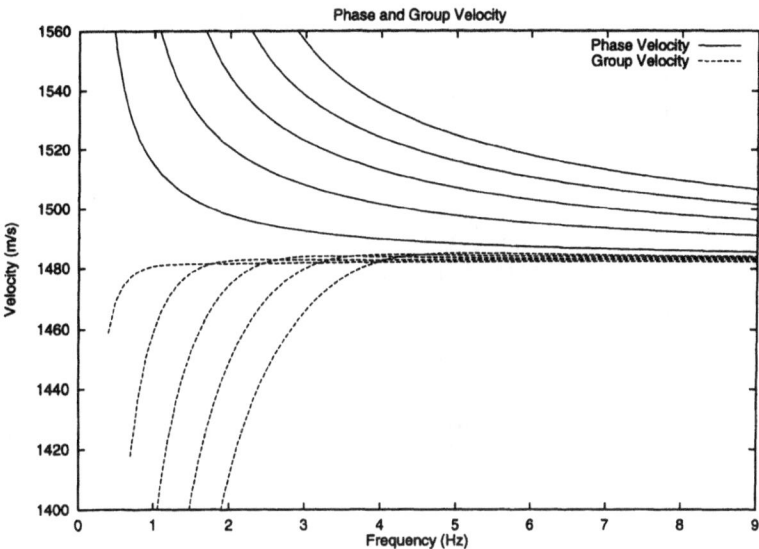

Figure 18
Phase and group velocity dispersion curves for the first 5 acoustic modes.

ocean. Because of this, we high-pass filter each mode at a frequency such that the waveform will fit within a 40 second time window.

The first 20 seconds of the waveforms at two depths are shown in Figure 19. At 1005 meters depth, the fundamental mode is substantially larger than the other modes, however at 2160 meters depth, the second mode is larger than the fundamental. The modes have been filtered in a band pass of 1.0 to 4.5 Hz, in addition to the mode-dependent high-pass filter described above.

Even though this is a two-dimensional problem, it requires a very large calculation because the distances of interest are large, about 100 km, the wave speed in the water is slow, and we are interested in frequencies higher than 1 Hz. The velocity model is the California coastal model discussed earlier and listed in Table 2. The ocean model is based on the bathymetric profile along the path to Loihi and through the Point Sur station (see Fig. 17), and the water velocity profile in the ocean near Point Sur (Fig. 15). We used a grid with a uniform spacing of 67 meters per grid cell, dimensions of 1500 × 500 grid blocks, and a time step of 0.005 seconds. The calculation was run for 20,000 cycles to obtain a duration of 100 seconds and required about 30 hours of CPU time on a DEC 2100 computer. The calculation should be accurate throughout the grid to at least 2.5 Hz. We expect to see numerical dispersion in the 2.5–4.5 Hz frequency band in the lower velocity parts of the grid.

Three calculations were performed, one with only the fundamental mode, one with the first 5 modes, and one with 5 modes in a structure with a constant ocean depth. This last calculation was done as a control experiment to check for numerical

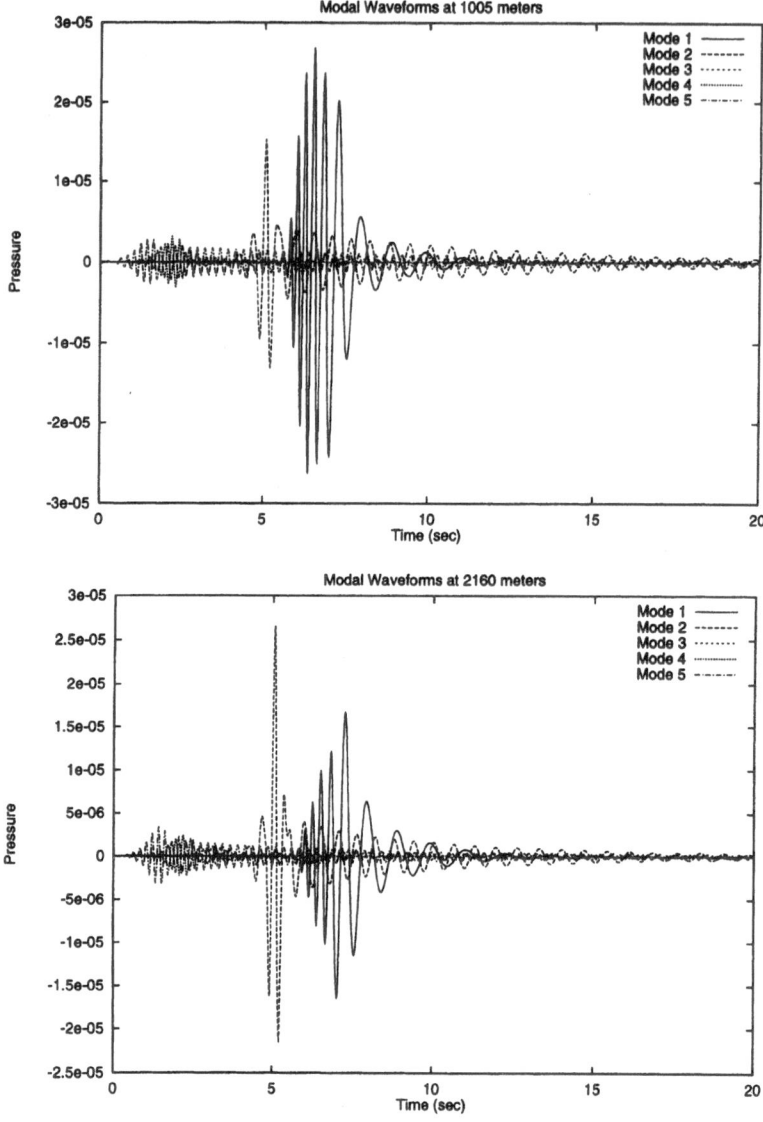

Figure 19
Input waveforms for first five modes at depths of 1005 and 2160 meters.

artifacts in the calculation. Velocities for the full grid were saved for each second of time in the calculation, and these were used to create images and animations of the velocity fields to help visualize the evolution of the velocity field. The results with a single mode and with five modes are very similar. There is a gradual decay of the hydroacoustic wave as it travels upslope, with body waves emitted continuously with varying amplitudes into the earth below, and a surface wave that gradually forms along the ocean bottom. At sharper bathymetric gradients the transmission is increased, and when the hydroacoustic wave reaches an ocean depth of 200 meters, there is a burst of energy much larger than anywhere else along the path. Strong surface waves are generated on land from the edge of the ocean to the boundary of the calculation. Figure 20 displays snapshots of the fields at

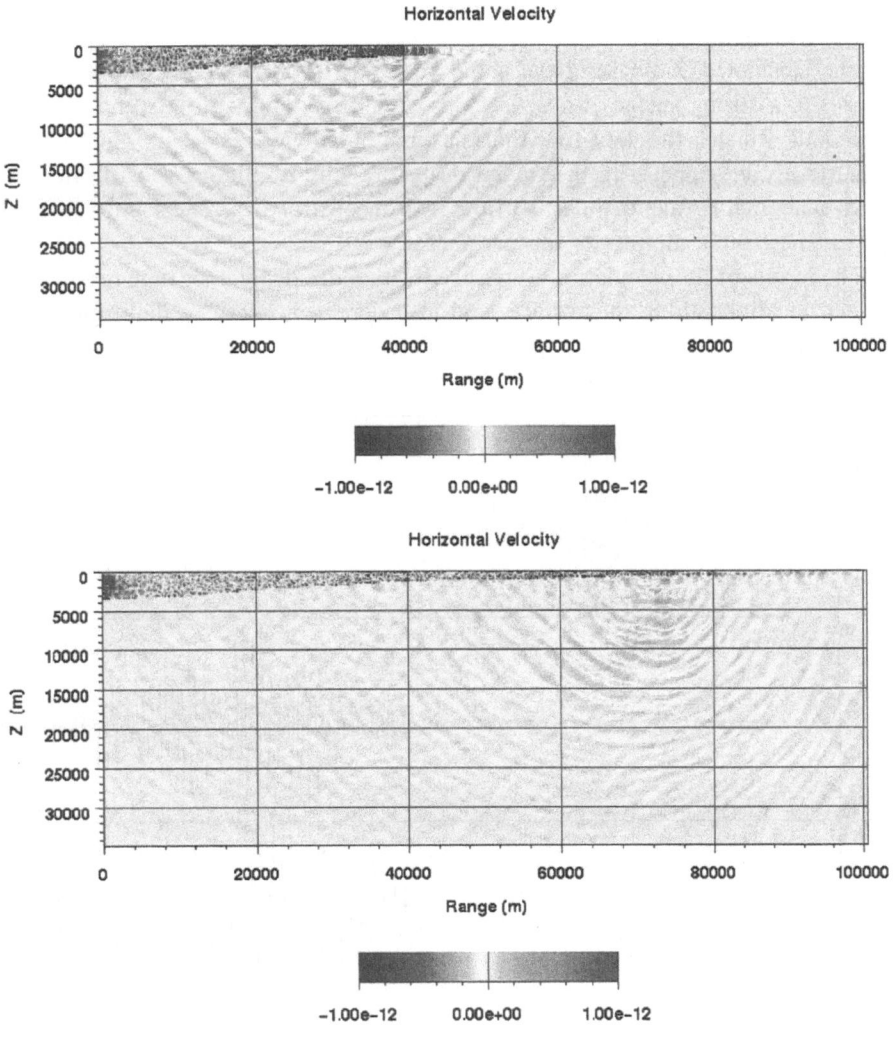

Figure 20
Snapshot of the calculation at 35 seconds (top) and 65 seconds (bottom). In the bottom figure, the dominant hydroacoustic arrival has reached a depth of 200 meters. At this point there is a burst of energy transferred as body waves from the ocean into land. Zero range corresponds to the point 3710 km from Loihi (see Fig. 18). The ocean/land interface at the surface is at location 87,368.

35 seconds and 65 seconds. The second figure indicates the burst of energy at 200 meter depth. Color animations of all three calculations can be viewed online at http://www.maxwell.com/products/geop/Movies/hydro.htm. The control calculation shows a small amount of energy transmission into the earth near the beginning of the calculation due to numerical inaccuracy in the finite source, although negligible transmission into the earth thereafter. This indicates that the numerical simulation of the modal solution propagating in the water is stable, and that the decay of the hydroacoustic wave is due to propagation across the ocean as it decreases in depth.

The calculated T phases have some very odd properties. As can be seen in Figure 20, a strong surface wave develops quite early and can be seen on land at 65 seconds, which is the same time that the burst of body waves occurs. Consequently the surface wave appears in the wavetrain before, and simultaneous with, the body waves generated by the final decay of the hydroacoustic wave. The T phase near the coast is therefore a mixture of seismic phases.

The results of the calculation were saved with a decidedly finer time resolution at selected locations along the surface, and at a depth of 740 m in the center of the sound channel. Figure 21 depicts the calculated horizontal velocity in the water at

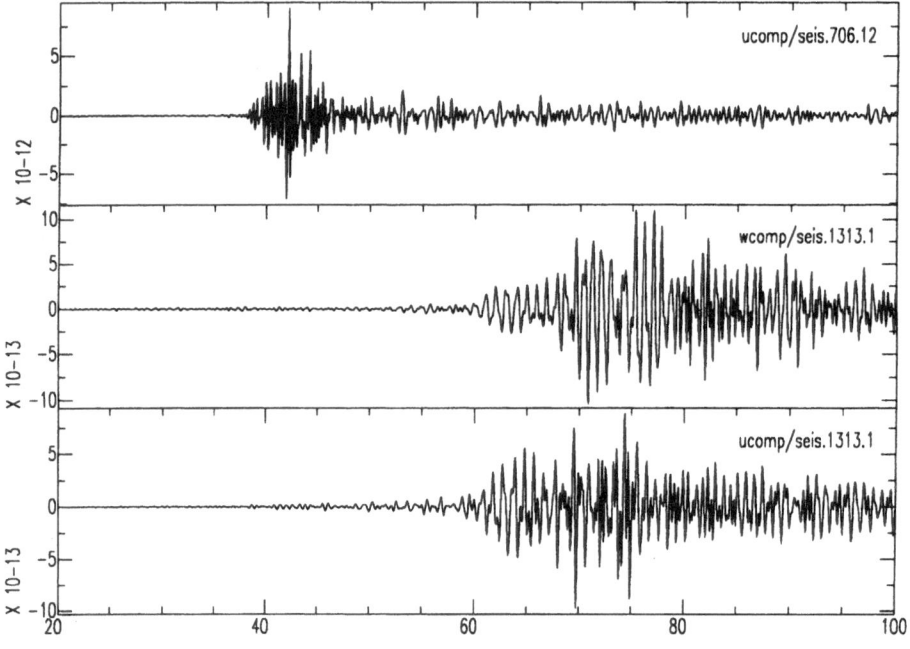

Figure 21

Calculated horizontal velocity at a depth of 740 meters in water at the location of the Point Sur station (top) and vertical (mid) and horizontal (bottom) components of the velocity on land close to the coast. Horizontal axis is time in seconds since the start of the calculation.

47,235 m, which corresponds approximately to the location of the Point Sur hydroacoustic station. Also shown are the horizontal and vertical components of the waveform on the surface at location 87,904 m, which is on land close to the coast. The amplitude ratios between the velocity on land and in the water are about 0.2, which is consistent with our earlier estimates of the upper bound of the transfer function. The waveforms on land are complex and longer in duration than the underwater waveform. From travel times and particle motion it is clear that the dominant energy is traveling at surface wave speeds. The complexity of the waveforms arises from the complex manner of wave generation over an extended region of the ocean bottom.

Figure 22 shows the spectral amplitude of the horizontal velocity in the 1–4 Hz frequency band underwater at the Point Sur location and the vertical velocity at a station on land close to the coast. The underwater spectrum is nearly flat across this frequency band. The spectra on land, however, exhibit a significant decline in amplitude with frequency, similar to the spectral decline observed for the coastal California stations. This suggests that during transmission from water to land, the higher frequencies are scattered more strongly than the lower frequencies, leading to a decline in high frequency content in the coastal waveforms.

Discussion

We have used *T*-phase observations in water and on land together with numerical calculations of *T*-phase propagation from water to land in order to understand the nature of *T*-phase conversion. The results are complex and at first glance contradictory. Whereas *T* phases are observed to travel at *P* wave speeds on land, the calculations reveal strong surface waves near the coast. However, the calculations also show lower amplitude body wave arrivals even very close to the coast, and strong body waves propagated away from conversion points along the ocean bottom. Some of this body wave energy will return farther inland as P_g and P_n phases. The high-frequency surface waves, on the other hand, can be expected to attenuate away very quickly and will not propagate to extended distances. We therefore expect to see strong surface waves and smaller *P*-wave arrivals near the coast, with the surface waves dying out and the *P* waves becoming dominant as the wave travels inland. The strong surface waves seen in this case may be somewhat anomalous, since they are formed in the ocean and propagate up the ocean bottom onto land. In a real earth with bottom sediments, these would be more strongly attenuated.

A similar study was performed by PISERCHIA *et al.* (1998), in which they modeled *T*-phase conversion from an explosive source in the ocean, observed on the islands of Mururoa and Fangataufa in the South Pacific Ocean. They used ray tracing instead of a modal solution, and calculated Green's functions along a vertical boundary in order to propagate the source onto the islands. They found that the *T* phase on land

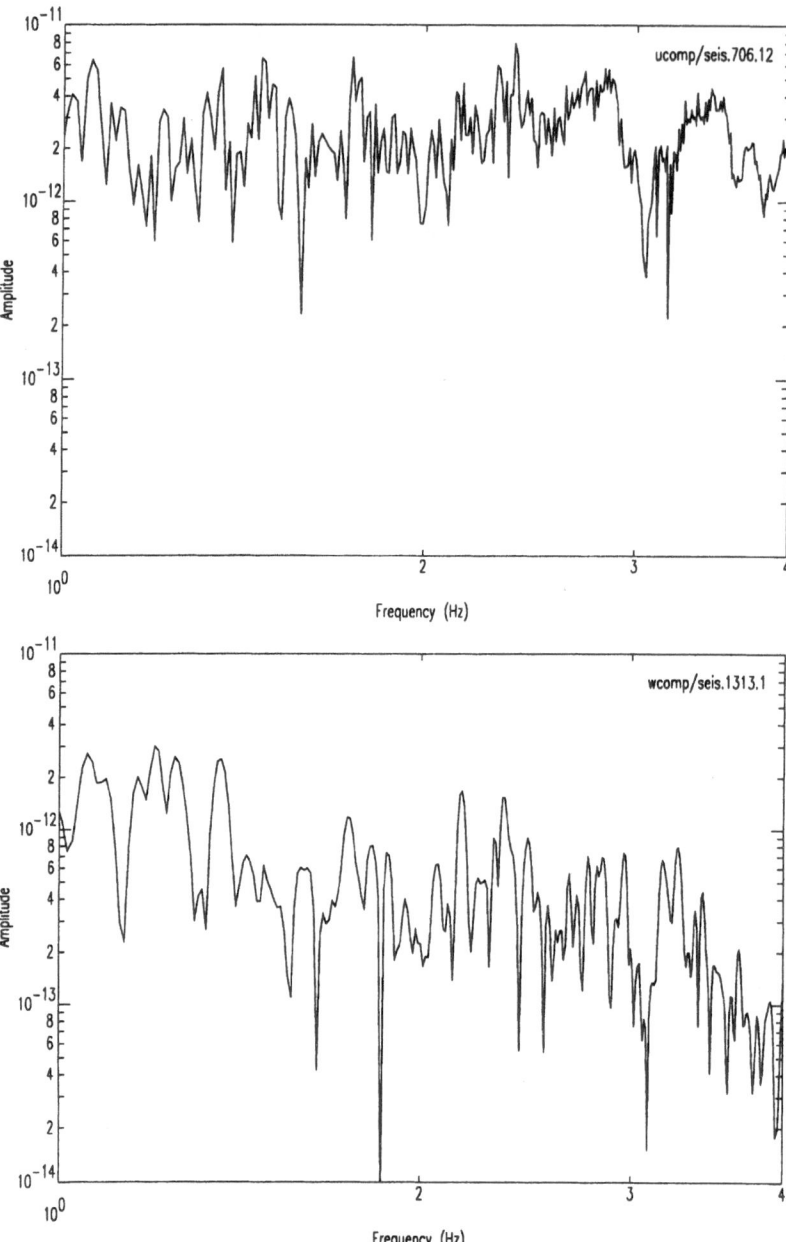

Figure 22
Calculated spectral amplitudes of the horizontal velocity at a depth of 740 meters underwater (top) and the
vertical velocity at the surface on land near the coast (bottom).

consisted of two *P* waves followed by two Rayleigh waves, where the multiple arrivals are identified as originating from different conversion points. The calculation was performed for a dominant frequency of 6 Hz, and in this case the calculated *P* and Rayleigh waves were found to be comparable in size. CANSI and BETHOUX (1985) modeled far inland *T* phases as converted *T-P* and *T-S* phases along a curve corresponding to a fixed depth in the ocean. They found good agreement with observed waveforms with synthetics composed only of *P* and *S* waves. DEGROOT-HEDLIN and ORCUTT (1997) modeled *T* phases by calculating the pressure on the ocean bottom from the fundamental mode propagating along a bathymetric profile and demonstrated that the shape and duration of the *T*-phase coda from Aleutian earthquakes could be modeled fairly well at Point Sur and Wake Island. TALANDIER and OKAL (1998) studied conversion of *T* phases on steep island slopes, using data from the Polynesian Seismic Network and found from the observations and ray-tracing arguments that the *T* phase consisted primarily of *P* waves at distances greater than 9 km from the conversion point. At closer distances, they found that the *T* phase was more complex and composed primarily of *S* waves and surface waves. They also suggested that only surface wave conversion would occur for slopes with angles of less than 16 degrees at short distances.

We have modeled propagation onto the California coast in considerable detail due to the availability of a good data set for comparison, and because we have stations both underwater and on land. The question arises as to how closely this models propagation onto the IMS *T*-phase stations which have different bathymetric profiles. Also, we would like to know how sensitive the calculations are to the details of the profile, and in particular to the slope of the ocean bottom on the approach to the coast. Figure 23 shows the bathymetric profiles off the coast of California in the direction of Loihi, passing through the Point Sur, RPV, and BKS stations. Also presented for comparison is the profile in the direction of Loihi off the coast of VIB. Although they differ in detail, the structures are comparable in general features. The slope off the coast of VIB is similar to the slope off BKS, although VIB is considerably closer to the coast. The slope through RPV is smaller. The slope used in the calculations is intermediate between these slopes (see Fig. 17). Note that there is considerable vertical exaggeration in these figures. Even the steepest part of the slope at VIB decreases by 1000 m over a distance of 10 km, which corresponds to an angle of about 6 degrees. The steepest slopes off the island *T*-phase stations change by 1000 m spanning a distance of four km, an angle of about 14 degrees. To assess the degree of difference this resulted in, we ran three additional finite difference calculations, using the same ocean structure and modes as in the calculations above, although with constant slopes of 10, 20, and 30 degrees. In each case, the coastal boundary was fixed at 80 km from the left edge of the grid. The resulting vertical displacements at a point 7 km inland from the coast are shown in Figure 24. The waveforms are remarkably similar to each other and to the earlier calculations. The waveforms

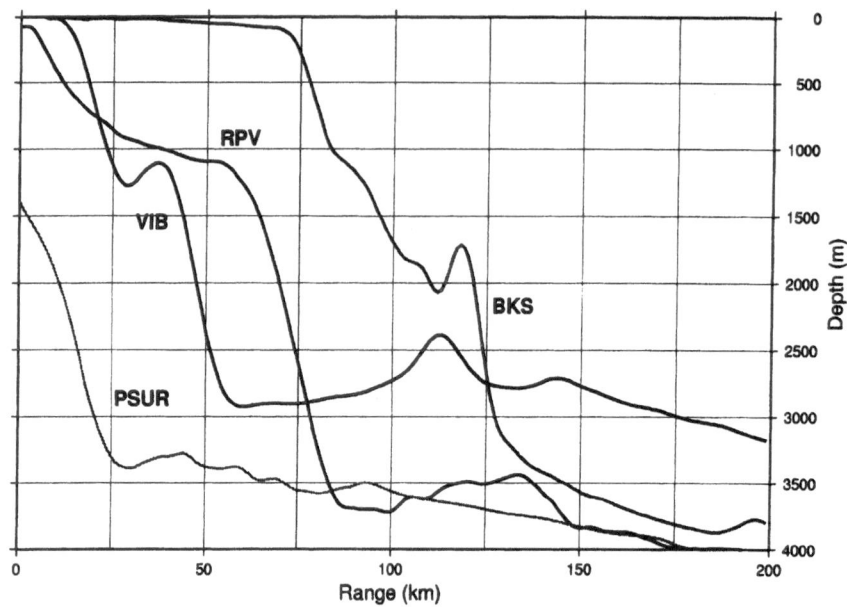

Figure 23
Great circle path showing bathymetry to PSUR, RPV, BKS, and VIB along the direction to Loihi from each station. Zero range corresponds to the station location. Vertical exaggeration makes the slopes appear steeper.

consist primarily of surface waves of comparable amplitude. The main difference is in the timing of the body wave, which close examination reveals arrives earlier than the surface wave for the 30 degree slope, and is within the surface wave wavetrain for the shallower slopes.

Conclusions

We have used observations of T phases underwater and on land to directly measure the transmission of energy from ocean to land, and have performed numerical simulations of T-phase propagation from ocean to land to obtain a better understanding of this process. The observations show that there is a significant decline in spectral amplitude with frequency on land, compared to observations in the ocean. Observation of waveforms from underwater explosions at VIB, however, shows that even with the high frequency degradation, sufficient high frequencies remain to identify such an event as an explosion. The observations also illustrate that T phases propagate primarily as P waves once they are far enough inland. The

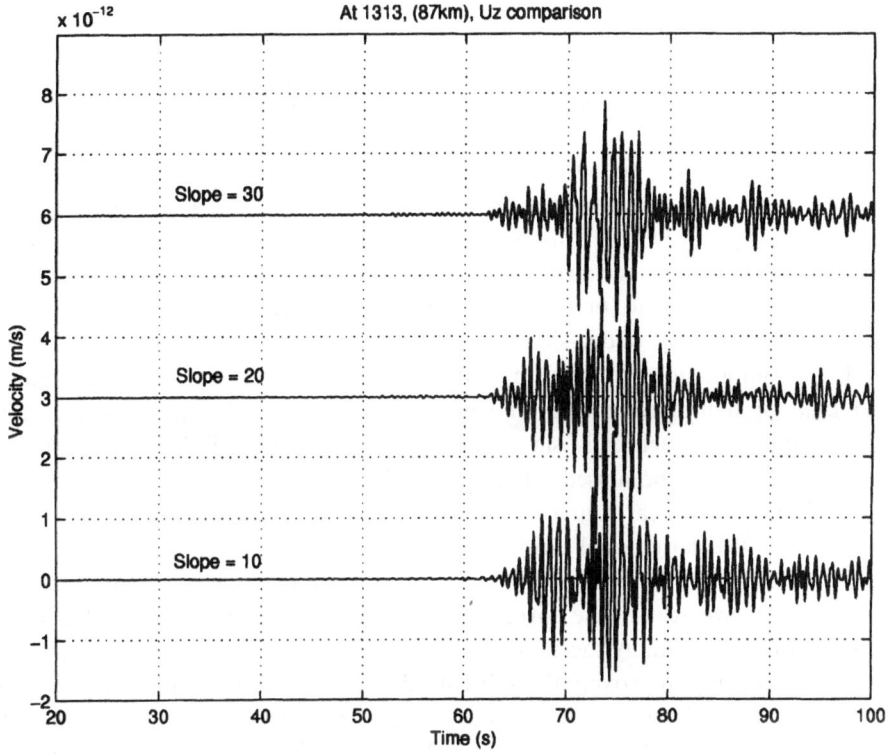

Figure 24

Vertical velocity waveforms at a point on the surface 7 km inland from the coast for the three calculations
with slopes of 10 (bottom), 20 (middle), and 30 (top) degrees.

numerical simulations provide considerable insight into the phenomena that occur
when hydroacoustic waves propagate onto land. The calculations reproduce the
spectral degradation observed in Coastal California stations. Both the calculations
and the observations show that *T* phases observed near the coast are composed
primarily of surface waves. Because the surface waves are generated over a more
extended region than body waves, they may arrive at near coastal stations earlier
than the body wave and obscure any body wave arrivals.

Acknowledgements

We thank Marcia McLaren of Pacific Gas and Electric for the use of her data and
for pointing out to us the large *T* phases from the Loihi events. We thank Heming Xu
for his assistance with the numerical calculations. This work was supported by
Defense Threat Reduction Agency contract DSWA01-97-C-0166.

Appendix A

Conversion of Modal Solution to Finite Difference Source Function

We have partitioned the simulation of the hydroacoustic phases into two separate computations. First we compute the incident hydroacoustic field from the deep ocean. This incident field is computed using normal mode superposition. A second computation simulates the evolution of the incident wavefield as it propagates through ocean-continent (or ocean-island) transition, represented by a 2-D cross section with a roughly wedge-shaped fluid layer overlying a solid elastic halfspace. The latter computation is done using a time-domain finite difference (TFD) method. This kind of hybrid approach, in which a semi-analytical solution (e.g., mode expansion) in a geometrically simple domain is linked to a TFD calculation in a complex domain, has been used extensively to simulate the propagation of explosion-generated, nonlinear near-source wavefields to teleseismic distances (e.g., DAY *et al.*, 1983, 1987). The latter application differs from the current one in that for the explosion source problem it is the transmitted field that has the semi-analytical representation, whereas in the current problem it is the incident field. Nonetheless, the technique is essentially the same.

To indicate the approximations involved, we start by describing an idealized version of the problem which can be partitioned exactly. Figure A1 shows the idealized problem schematically. A surface Σ' encloses a volume V with an arbitrarily complex distribution of elastic properties. Region V is embedded in a horizontally stratified halfspace HS. The unit outward normal to Σ' is denoted by v. We denote by Σ that part of Σ' which does not coincide with the free surface of HS. An incident hydroacoustic field is computed for HS, and we denote its displacement by $\mathbf{u}^{inc}$.

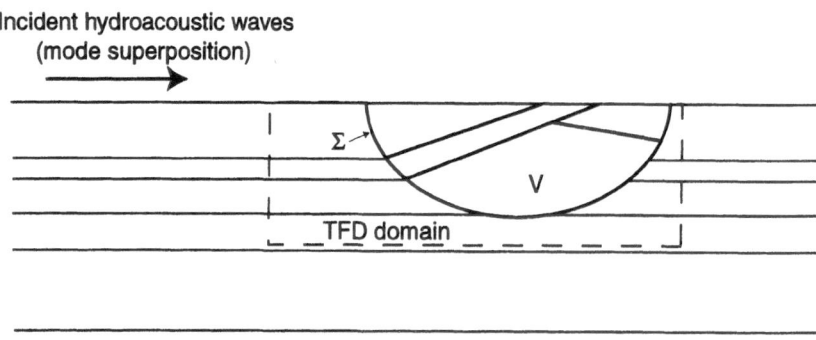

Figure A1
Schematic drawing of the idealized problem, in which V is superimposed on a laterally homogeneous halfspace HS.

Since all sources are external to region V, the (total) displacement field **u** in V (assuming initial quiescence) can be written as a surface integral of the displacements and tractions on Σ,

$$u_i(\mathbf{x}, t) = \int_{\Sigma} d\Sigma C_{jkpq}(\xi)[u_{p,q}(\xi, t) * G_{ij}(\mathbf{x}, \xi, t) - u_j(\xi, t) * G_{ip,q}(\mathbf{x}, \xi, t)]v_k \qquad (1)$$

(e.g., AKI and RICHARDS, 1980), where **C** is the elastic modulus tensor, **G** is the elastodynamic Green's tensor for the full, laterally varying, halfspace model, and * denotes convolution. The integral in (1) need only be taken over Σ, rather than over the closed surface Σ', because both terms in (1) vanish at the free surface. Next, we write the total displacement field **u** as the sum of the incident field $\mathbf{u}^{inc}$ and a scattered field $\mathbf{u}^s$. That is, we define the scattered field as the difference between the total and incident field, $\mathbf{u} - \mathbf{u}^{inc}$. Then (1) breaks into separate integrals involving $\mathbf{u}^{inc}$ and $\mathbf{u}^s$, respectively. It is easy to show that the integral containing $\mathbf{u}^s$ vanishes (noting that $\mathbf{u}^s$ on and exterior to Σ can be viewed as a solution to the equations of motion for the flat-layered model HS, with excitation by equivalent sources wholly contained in V). Therefore, (1) can be rewritten in terms of the incident field only,

$$u_i(\mathbf{x}, t) = \int_{\Sigma} d\Sigma C_{jkpq}(\xi)[u_{p,q}^{inc}(\xi, t) * G_{ij}(\mathbf{x}, \xi, t) - u_j^{inc}(\xi, t) * G_{ip,q}(\mathbf{x}, \xi, t)]v_k . \qquad (2)$$

Equation (2) can be written formally as a volume integral, by means of the Dirac delta distribution (e.g., BURRIDGE and KNOPOFF, 1964; BACKUS and MULCAHEY, 1976),

$$u_i(\mathbf{x}, t) = \int_V dV_\eta[f_j(\eta, t) * G_{ij}(\mathbf{x}, \eta, t) + m_{pq}(\eta, t) * G_{ip,q}(x, \eta, t)] , \qquad (3)$$

where

$$f_j(\eta, t) \equiv \int_{\Sigma} d\Sigma C_{jkpq} u_{p,q}^{inc}(\xi, t)v_k\delta(\eta, \xi) \qquad (4)$$

and

$$m_{pq}(\eta, t) \equiv - \int_{\Sigma} d\Sigma C_{jkpq} u_j^{inc}(\xi, t)v_k\delta(\eta, \xi) \qquad (5)$$

where $\delta(\eta, \xi)$ is the product of Dirac delta functions in the coordinates, $\delta(\eta_1 - \xi_1)\delta(\eta_2 - \xi_2)\delta(\eta_3 - \xi_3)$. In volume integral (3), **f** can be interpreted as an equivalent body force distribution per unit volume, and **m** can be interpreted as a moment tensor density distribution per unit volume. The equation of motion is then

$$\rho\ddot{u}_i - (C_{ijkl}u_{k,l} - m_{ij})_{,j} - f_i = 0 , \qquad (6)$$

where ρ is the mass density.

In the 2-D (plane strain) problem in x_1, x_2, the surface Σ is replaced by a curve C in the x_1, x_2 plane. The source terms (4) and (5) become

$$f_j(\eta, t) \equiv \int_C dl\, C_{jkpq} u_{p,q}^{inc}(\xi, t) v_k \delta(\eta, \xi) \tag{7}$$

and

$$m_{pq}(\eta, t) \equiv - \int_C dl\, C_{jkpq} u_j^{inc}(\xi, t) v_k \delta(\eta, \xi) \ . \tag{8}$$

Source terms (7, 8) provide an exact connection between the incident wavefield and the 2-D TFD computations. We will derive a discrete form of these 2-D source terms for the TFD computations, after we have made simplifying approximations.

Our application of (7, 8) is not quite exact, even apart from the approximations involved in discretizing (3). Initially, we apply absorbing boundary conditions (the first-order conditions of CLAYTON and ENQUIST, 1977) to the side and bottom boundary of the TFD domain. These boundaries produce small, but spurious, reflections of incident energy, as is well known. Secondly, we must deal with a model slightly different from the ideal one depicted in Figure A1. A configuration more representative of our problem is shown schematically in Figure A2. Now the region inside which the structure departs from HS is not finite in extent. Instead, it extends

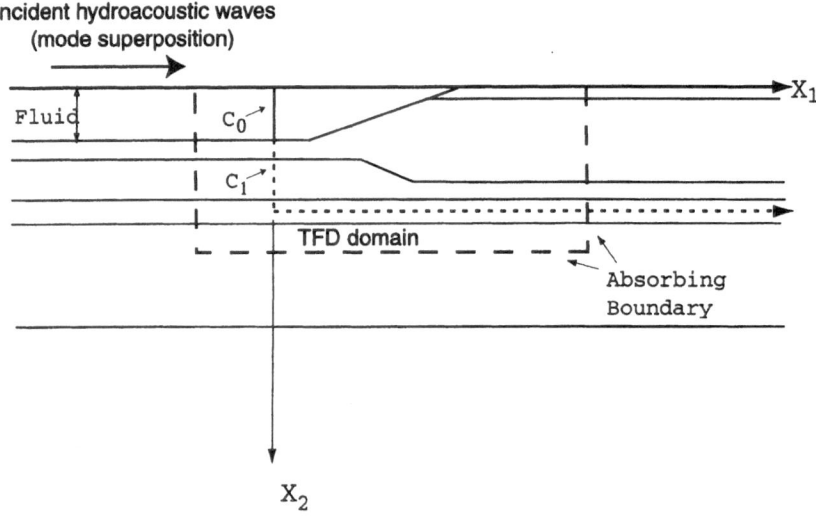

Figure A2

The 2-D hydroacoustic problem. Boundary C encloses all portions of the problem domain that depart from the laterally homogeneous halfspace HS for which the modal solution is computed. C_0 is that part of the boundary which lies in the fluid part of HS.

infinitely to the right. For the above formulation to be exact, the boundary C (on which the incident-field equivalent sources are applied) would have to extend sufficiently far to the right that the P-wave propagation time to its terminus exceeds the duration of the TFD computation. However, we will assume that the incident acoustic field has much higher amplitude in the fluid layers than in the solid earth. On that basis we neglect equivalent source contributions from the part of C in the solid earth, denoted C_1, and apply the source only to the fluid portion, C_0. On the basis of numerical experiments, we believe this simplification has much less effect on the solution than does the use of absorbing boundaries. For convenience, we take C_0 to be vertical (which entails no further approximation).

The numerical implementation of the new source terms is very simple. The moment density components in (3) enter the equations exactly as the elastic stress components $C_{ijkl}u_{k,l}$. Therefore in the TFD algorithm we replace $C_{ijkl}u_{k,l}$ at each time increment with $C_{ijkl}u_{k,l} - \mathbf{m}$, and this combination is spatially differenced exactly as the components of $C_{ijkl}u_{k,l}$ are spatially differenced in the homogeneous equations. Likewise, the components of $\mathbf{f}$ enter exactly as do the acceleration components, and the components of $\mathbf{f}$ are added directly to the acceleration.

It remains to discretize $\mathbf{m}$ and $\mathbf{f}$. Figure A3 shows the geometry of the TFD grid cells through which C_0 passes. We align C_0 with the stress points (at grid cell centers). To develop a discrete representation $\bar{\mathbf{m}}$ of $\mathbf{m}$ at a stress point, we evaluate the incident field at that point, denoting its value by $\bar{\mathbf{u}}^{\text{inc}}$, and make the approximation that $\mathbf{u}^{\text{inc}}$ is constant at this value throughout the grid cell. Then we apply (8) to evaluate $\mathbf{m}$ throughout the cell, using this approximation for $\mathbf{u}^{\text{inc}}$, and define $\bar{\mathbf{m}}$ as the mean value taken by this estimate of $\mathbf{m}$ in that grid cell. The discrete force densities $\bar{\mathbf{f}}$ are determined the same way. The results are

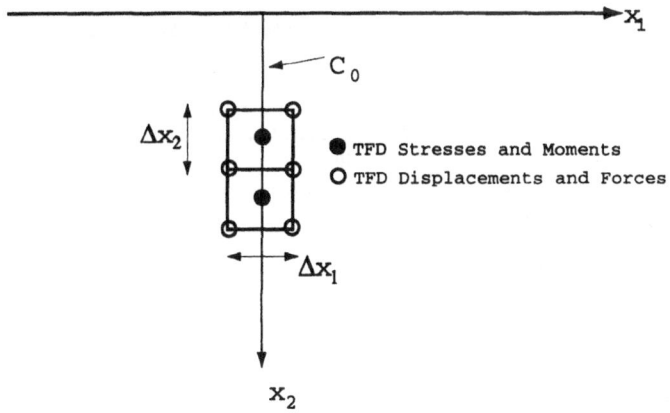

Figure A3

Relationship of the TFD grid to C_0. Stresses (and moments) are centered at points which lie on C_0, while displacements (and forces) are centered at one half cell dimension to the right and left of C_0.

$$\bar{f}_j = C_{jkpq}\bar{u}_{p,q}^{\text{inc}}v_k\Delta x_1^{-1} \tag{9}$$

and

$$\bar{m}_{pq} = -C_{jkpq}\bar{u}_j^{\text{inc}}v_k\Delta x_1^{-1} \ . \tag{10}$$

Finally, with our choice of a vertical C_0 confined to the fluid, (9) and (10) reduce to

$$\bar{f}_1 = \bar{p}^{\text{inc}}\Delta x_1^{-1} \ , \tag{11}$$

$$\bar{m}_{11} = \bar{m}_{22} = \kappa\bar{u}_1^{\text{inc}}\Delta x_1^{-1} \ , \tag{12}$$

where $\bar{p}^{\text{inc}}$ is the incident pressure field at the grid cell center, and all other source components are zero.

An additional modification results from the fact that force densities must be applied at the displacement points rather than the stress points (Fig. A3). After calculating $\bar{f}$ at a stress point (via Equation (11)), we distribute 1/4 its value onto each of the 4 adjacent displacement points. Each displacement point receives contributions from the cell above and cell below in this process.

References

ACHENBACH, *Wave Propagation in Elastic Solids* (North-Holland Publishing Company, New York, 1973).

AKI, K., and RICHARDS, P. G., *Quantitative Seismology*, vol. I (W. H. Freeman and Co., San Francisco, 1980).

BACKUS, G., and MULCAHY, M. (1976), *Moment Tensors and other Phenomenological Descriptions of Seismic Sources – II. Discontinuous Displacements*, Geophys. J. R. astro. Soc. 47, 301–329.

BISWAS, N. N., and KNOPOFF, L. (1970), *Exact Earth-flattening Calculation for Love Waves*, Bull. Seismol. Soc. Am. 60, 1123–1137.

BURRIDGE, R., and KNOPOFF, L. (1964), *Body Force Equivalents for Seismic Dislocations*, Bull. Seismol. Soc. Am. 54, 1875–1888.

CANSI, Y., and BETHOUX, N. (1985), *T Waves with Long Inland Paths: Synthetic Seismograms*, J. Geophys. Res. 90, 5459–5465.

CAPLAN-AUERBACH, J., DUENNEBIER, F. K., OKUBO, P., and KONG, L. (1999), *Seismicity and Velocity Structure of the Loihi Seamount from the 1996 Earthquake Swarm*, in preparation.

CLAYTON, R., and ENGQUIST, B. (1977), *Absorbing Boundary Conditions for Acoustic and Elastic Wave Equations*, Bull. Seismol. Soc. Am. 67, 1529–1540.

COOK, R. W., and STEVENS, J. L. (1998), *TP Phase Observations at the PIDC*, Transact. Am. Geophys. Union 79, P. F558, November.

DAY, S. M., RIMER, N., and CHERRY, J. T. (1983), *Surface Waves from Underground Explosions with Spall: Analysis of Elastic and Nonlinear Source Models*, Bull. Seismol. Soc. Am. 73, 247–264.

DAY, S. M., CHERRY, J. T., RIMER, N., and STEVENS, J. L. (1987), *Nonlinear Model of Tectonic Release from Underground Explosions*, Bull. Seismol. Soc. Am. 77, 996–1016.

DE GROOT-HEDLIN, C., and ORCUTT, J. (1997), *Observations of T Phases from Pacific Earthquake Events: Implications for Seismic/Acoustic Coupling*, Final report to Phillips Laboratory PL-TR-97-2144, November.

D'SPAIN, G. L., BERGER, L. P., KUPERMAN, W. A., STEVENS, J. L., and BAKER, G. E. (2001), *Normal Mode Composition of Earthquake T Phases*, Pure appl. geophys. 158, 475–512

JENSEN, F. B., KUPERMAN, W. A., PORTER, M. B., and SCHMIDT, H., *Computational Ocean Acoustics* (American Institute of Physics Press, New York, 1993).

McLaughlin, K. L., and Day, S. M. (1994), *3D Elastic Finite-difference Seismic-wave Simulations*, Computers in Physics *8* (6), 656–663.

Mooney, W. D., Laske, G., and Guy Masters, T. (1998), Crust 5.1: A Global Crustal Model at 5° × 5°, J. Geophys. Res. *103*, 727–747.

NOAA (1998), *5-Min Global Digital Terrain Models*, Web site of the National Geophysical Data Center, http://www.ngdc.noaa.gov/seg/fliers/se-1104.html.

Oliver, J., and Ewing, M. (1958), *Short-period Oceanic Surface Waves of the Rayleigh and First Shear Modes*, Transact. Am. Geophys. Union *30*, 482–485.

Piserchia, P. F., Virieux, J., Rodrigues, D., Gaffet, S., and Talandier, J. (1998), *Hybrid Numerical Modelling of T-wave Propagation: Application to the Midplate Experiment*, Geophys. J. Int. *133*, 789–800.

Scripps (1998), Web site of the NEMO Oceanographic data server, Scripps Institution of Oceanography, National Oceanographic Data Center's Station Data File http://nemo.ucsd.edu/hs.html

Stevens, J. L., Baker, G. E., Murphy, J. R., Cook, R. W., D'Spain, G., Berger, L. P., and Khristoforov, B. D. (1998), *T-phase Excitation and Transfer Function Research*, Proceedings of the 20th Seismic Research Symposium on Monitoring a Comprehensive Test-Ban Treaty, 21–23 September.

Stevens, J. L., and McLaughlin, K. L. (1997), *Improved Methods for Regionalized Surface Wave Analysis*, Maxwell Technologies Final Report submitted to Phillips Laboratory, MFD-TR-97-15887, September.

Talandier, J., and Okal, E. A. (1998), *On the Mechanism of Conversion of Seismic Waves to and from T Waves in the Vicinity of Island Shores*, Bull. Seismol. Soc. Am. *88*, 621–632.

University of Hawaii (1998), Web site of the Hawaii Center for Volcanology, School of Ocean and Earth Science and Technology, University of Hawaii, http://www.soest.hawaii.edu.

(Received June 29, 1999, revised December 17, 1999, accepted January 3, 2000)

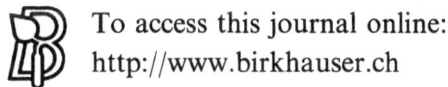

To access this journal online:
http://www.birkhauser.ch

Pure appl. geophys. 158 (2001) 567–603
0033–4533/01/030567–37 $ 1.50 + 0.20/0

❙Pure and Applied Geophysics

Identification Criteria for Sources of T Waves Recorded in French Polynesia

JACQUES TALANDIER[1] and EMILE A. OKAL[2]

Abstract—From a data set of 150 digital records of T phases from 71 sources obtained on seismometers of the Polynesian Seismic Network, we define a discriminant separating earthquake and explosion sources, which uses the maximum amplitude of recorded ground velocity, measured on its envelope, e_{Max} (in µm/s), and the duration of the phase measured at 1/3 of maximum amplitude, $\tau_{1/3}$ (in seconds). Earthquake sources and man-made explosions are effectively separated in a log-log space by the straight line

$$\log_{10} e_{Max} = 4.9 \log_{10} \tau_{1/3} - 4.1 \ .$$

Other criteria in both the time and frequency domains fail to reliably separate the populations of the various kinds of events. The application of this technique to analog records of large-scale man-made explosions carried out in the 1960s confirms that it provides an adequate discriminant over 3.5 orders of magnitude of ground velocity.

Key words: Seismic discrimination, hydroacoustics, T phases.

1. Introduction and Background

The Comprehensive Nuclear-Test-Ban Treaty (CTBT) has mandated as part of its International Monitoring System (IMS) the deployment of so-called T-phase stations, which consist of high-frequency seismometers located in the immediate vicinity of coastlines, for the purpose of recording seismic waves converted from acoustic energy propagated in the SOFAR channel of the world's oceans. The advantage of this design over more traditional hydrophone stations (also mandated by the CTBT) stems from a considerable simplification of logistics for powering, maintenance and data retrieval. It builds on the longstanding observation that the exceptionally efficient propagation of T waves in the water mass of the ocean allows the detection and location of very small, very remote sources in the marine

---
[1] Département Analyse et Surveillance de l'Environnement, Commissariat à l'Energie Atomique Boîte Postale 12, 91680 Bruyères-le-Châtel, France.
[2] Department of Geological Sciences, Northwestern University, Evanston, IL 60201, U.S.A. E-mail: emile@earth.nwu.edu

environment, all from land-based instruments. Indeed, the first positive identifications of T phases were made on seismometers (LINEHAN, 1940; RAVET, 1940).

The routine operation of hydrophone networks over several decades in the Pacific
Basin has long established that T phases can vastly increase detection capabilities in
the basin (e.g., DUENNEBIER and JOHNSON, 1967), with similar observations at land-
based networks, such as the Polynesian Seismic Network (Réseau Sismique
Polynésien; hereafter RSP) headquartered in Papeete, Tahiti. Thus, the operational
framework for T-phase and hydrophone stations in the IMS is that of the detection
and location of an event based entirely on T phases, without the benefit of additional
seismic phases, such as P waves. It is also in this context that a further step must be
performed, i.e., the identification of the nature of the source as either a natural
phenomenon or a man-made explosion.

The purpose of the present paper is to build on a data set of T phases recorded
at the RSP over 35 years, both from earthquakes (documented and located from
their seismic waves), and from known marine explosions, and to explore
systematically a number of criteria in the quest for a reliable discriminant between
natural and man-made sources. We show that a comparison between maximum
amplitude (determined on a smoothed envelope) and duration at 1/3 of maximum
provides a reliable means of separating earthquakes from explosions, which can be
applied over 3.5 orders of magnitude of ground velocity signal. We find it
imperative, however, to use stations located on atolls which feature a simpler and
more efficient acoustic-to-seismic conversion than do volcanic high island sites
(TALANDIER and OKAL, 1998).

2. Methodology

Data Set

Table 1 lists the digital data set used in this study, comprising 150 seismograms of
T phases from 71 events recorded at short-period seismic stations of the Polynesian
Seismic Network (Réseau Sismique Polynésien, hereafter RSP) during the period
1971–1997. This data set is complemented by records from earlier events (1962–1970)
of generally much larger amplitudes, and for which the data processing had to be
adapted to the analog records available at those dates (Table 2). For this reason,
these much larger sources are the subject of a separate discussion in Section 4.

The RSP network, described most recently by TALANDIER (1993), comprises 17
permanent short-period stations. During local refraction campaigns, a number of
temporary sites were instrumented. We distinguish in this study between stations
located on atolls ("(A)" in Tables 1 and 2) and those on high volcanic islands
("(H)"): we have shown in TALANDIER and OKAL (1998) that the conditions of
conversion of seismic energy to and from acoustic (T) energy differ fundamentally for

Table 1

Digital T-wave records used in this study

Date		Origin Time	Station		Nature	CODE NAME (E) or Region	Event size	Reference
D M (J) Y		GMT	Code	Site	(†)	(E, S, H, V)	(*)	(§)
9 OCT	(282) 1971	18:14	RUV	(A)	E	Polynesia	0.082 t	
9 OCT	(282) 1971	18:14	TPT	(A)	E			
30 SEP	(273) 1979	12:45	VIV	(A)	V	Macdonald		a
21 AUG	(234) 1980	15:00	DIN	(A)	E	Vancouver Is.		
21 AUG	(234) 1980	15:00	DOR	(A)	E			
21 AUG	(234) 1980	15:00	IRN	(A)	E			
21 AUG	(234) 1980	15:00	RUV	(A)	E			
21 AUG	(234) 1980	15:00	TPT	(A)	E			
10 NOV	(315) 1980	11:26	VIV	(A)	V	Macdonald		a
24 DEC	(359) 1980	16:09	AFR	(H)	V			
24 DEC	(359) 1980	16:09	VIV	(A)	V			
15 FEB	(046) 1981	16:18	VIV	(A)	V			
8 MAR	(067) 1981	13:59	VAH	(A)	H	Mehetia	3.1 M_L	b
8 MAR	(067) 1981	14:30	VAH	(A)	H	Mehetia	3.3 M_L	b
20 MAR	(079) 1981	15:00	VAH	(A)	H	Mehetia	3.1 M_L	b
16 DEC	(350) 1981	00:43	VAH	(A)	E	Polynesia	0.082 t	c
11 FEB	(042) 1982	18:54	VAH	(A)	E	Polynesia	0.082 t	c
11 FEB	(042) 1982	19:15	VAH	(A)	E			
11 FEB	(042) 1982	22:04	VAH	(A)	E			
1 MAR	(060) 1982	22:37	VIV	(A)	V	Macdonald		d
25 MAR	(084) 1982	18:02	VAH	(A)	E	Polynesia	0.082 t	c
25 MAR	(084) 1982	18:24	VAH	(A)	E	Polynesia	0.082 t	c
25 MAR	(084) 1982	19:58	FGA	(A)	E	Polynesia	0.082 t	c
25 MAR	(084) 1982	20:20	FGA	(A)	E	Polynesia	0.082 t	c
14 MAR	(073) 1983	19:15	VIV	(A)	V	Macdonald		d
24 MAR	(083) 1985	17:05	DOR	(A)	E	PSPM	0.9 t	e
24 MAR	(083) 1985	17:05	FGA	(A)	E			
24 MAR	(083) 1985	17:05	IRN	(A)	E			
24 MAR	(083) 1985	22:04	DOR	(A)	E	PSPM	0.9 t	e
24 MAR	(083) 1985	22:04	FGA	(A)	E			
24 MAR	(083) 1985	22:04	IRN	(A)	E			
26 MAR	(085) 1985	16:03	DOR	(A)	E	PSPM	1.0 t	e
26 MAR	(085) 1985	16:03	FGA	(A)	E			
26 MAR	(085) 1985	16:03	IRN	(A)	E			
26 MAR	(085) 1985	16:03	TPT	(A)	E			
26 MAR	(085) 1985	22:05	DOR	(A)	E	PSPM	1.4 t	e
26 MAR	(085) 1985	22:05	FGA	(A)	E			
26 MAR	(085) 1985	22:05	IRN	(A)	E			
26 MAR	(085) 1985	22:05	TPT	(A)	E			
30 MAR	(089) 1985	22:01	TPT	(A)	E	PSPM	0.8 t	e
31 MAR	(090) 1985	15:33	DOR	(A)	E	PSPM	0.8 t	e
31 MAR	(090) 1985	15:33	FGA	(A)	E			
31 MAR	(090) 1985	15:33	IRN	(A)	E			
31 MAR	(090) 1985	15:33	TPT	(A)	E			
1 APR	(091) 1985	18:31	DOR	(A)	E	PSPM	0.5 t	e
1 APR	(091) 1985	18:31	FGA	(A)	E			
1 APR	(091) 1985	18:31	IRN	(A)	E			

Table 1

(*Continued*)

Date	Origin Time	Station		Nature	CODE NAME (E) or Region	Event size	Reference
D M (J) Y	GMT	Code	Site	(†)	(E, S, H, V)	(*)	(§)
1 APR (091) 1985	18:31	TPT	(A)	E			
1 APR (091) 1985	29:31	TPT	(A)	E	PSPM	0.2 t	e
30 MAY (150) 1985	01:22	PMO	(A)	E			
31 MAY (151) 1985	05:42	PMO	(A)	M			
18 AUG (230) 1986	03:37	PMO	(A)	M			
18 AUG (230) 1986	03:38	PMO	(A)	M			
19 AUG (231) 1986	03:59	PMO	(A)	M			
3 SEP (246) 1986	06:04	PMO	(A)	M			
3 SEP (246) 1986	06:05	PMO	(A)	M			
4 SEP (247) 1986	12:38	TPT	(A)	E			
17 SEP (260) 1986	23:29	RUV	(A)	E			
17 SEP (260) 1986	23:29	TPT	(A)	E			
9 DEC (343) 1989	10:00	AFR	(H)	E	MIDPLATE	0.3 t	f
9 DEC (343) 1989	10:00	HUA	(A)	E			
9 DEC (343) 1989	10:00	MEH	(H)	E			
9 DEC (343) 1989	10:00	PAE	(H)	E			
9 DEC (343) 1989	10:00	PPN	(H)	E			
9 DEC (343) 1989	10:00	PPT	(H)	E			
9 DEC (343) 1989	10:00	TIA	(H)	E			
9 DEC (343) 1989	10:00	TVO	(H)	E			
9 DEC (343) 1989	17:00	PPN	(H)	E	MIDPLATE	0.3 t	f
9 DEC (343) 1989	17:00	TIA	(H)	E			
9 DEC (343) 1989	17:00	TVO	(H)	E			
9 DEC (343) 1989	18:00	FGA	(A)	E	MIDPLATE	0.3 t	f
13 DEC (347) 1989	19:45	VAH	(A)	E	MIDPLATE	0.025 t	f
15 DEC (349) 1989	02:00	DIN	(A)	E	MIDPLATE	0.3 t	f
15 DEC (349) 1989	02:00	FGA	(A)	E			
15 DEC (349) 1989	02:00	VAH	(A)	E			
21 DEC (355) 1989	22:42	PMO	(A)	V	Mariana Is.		
21 DEC (355) 1989	23:34	PMO	(A)	V	Mariana Is.		
22 DEC (356) 1989	16:00	MKT	(A)	E	MIDPLATE	1.0 t	f
22 DEC (356) 1989	18:00	MKT	(A)	E	MIDPLATE	0.3 t	f
23 DEC (357) 1989	07:37	PMO	(A)	V	Mariana Is.		
28 DEC (362) 1989	16:00	MEH	(H)	E	MIDPLATE	0.2 t	f
6 MAY (126) 1993	17:39	PMO	(A)	V	Mariana Is.		
8 JUN (159) 1993	12:57	PMO	(A)	H	Hawaii	5.2 m_b	g
8 JUN (159) 1993	12:57	RUV	(A)	H			
8 JUN (159) 1993	12:57	TIA	(H)	H			
8 JUN (159) 1993	12:57	TPT	(A)	H			
27 JUN (178) 1994	18:30	RUV	(A)	E	California		
27 JUN (178) 1994	18:30	TPT	(A)	E			
6 JAN (006) 1995	22:37	PPT	(H)	S	Hokkaido	3.3×10^{26} dyn-cm	h
6 JAN (006) 1995	22:37	TIA	(H)	S			
19 FEB (050) 1995	04:03	PPT	(H)	S	California	9.9×10^{25} dyn-cm	h
19 FEB (050) 1995	04:03	RUV	(A)	S			
19 FEB (050) 1995	04:03	TIA	(H)	S			
19 FEB (050) 1995	04:03	TPT	(A)	S			

Table 1

(*Continued*)

Date D M (J) Y	Origin Time GMT	Station Code	Station Site	Nature (†)	CODE NAME (E) or Region (E, S, H, V)	Event size (*)	Reference (§)
9 MAR (068) 1995	08:12	TIA	(H)	H	Hawaii	3.9 m_b	g
14 MAR (073) 1995	17:33	AFR	(H)	S	Alaska	2.2×10^{25} dyn-cm	h
14 MAR (073) 1995	17:33	PMO	(A)	S			
14 MAR (073) 1995	17:33	PPT	(H)	S			
14 MAR (073) 1995	17:33	RUV	(A)	S			
14 MAR (073) 1995	17:33	TIA	(H)	S			
14 MAR (073) 1995	17:33	TPT	(A)	S			
1 NOV (305) 1995	00:35	MEH	(H)	S	Chile	1.14×10^{26} dyn-cm	i
1 NOV (305) 1995	00:35	TET	(A)	S			
1 NOV (305) 1995	00:35	TIA	(H)	S			
16 JAN (016) 1996	20:46	TVO	(A)	S			
16 JAN (016) 1996	20:46	VAH	(A)	S			
7 FEB (038) 1996	21:36	PMO	(A)	S	Kuril Is.	6.4×10^{26} dyn-cm	j
7 FEB (038) 1996	21:36	PPT	(H)	S			
7 FEB (038) 1996	21:36	TET	(A)	S			
7 FEB (038) 1996	21:36	TIA	(H)	S			
30 MAR (090) 1996	09:56	AFR	(H)	S	Eltanin F.Z.	1.5×10^{25} dyn-cm	
30 MAR (090) 1996	09:56	VAH	(A)	S			
30 JUN (182) 1996	22:27	AFR	(H)	S	Eltanin F.Z.	1.7×10^{24} dyn-cm	k
30 JUN (182) 1996	22:27	VAH	(A)	S			
22 JUL (204) 1996	10:33	PPT	(H)	H	Hawaii	3.7 M_D	g
22 JUL (204) 1996	10:33	RUV	(A)	H			
22 JUL (204) 1996	10:33	TIA	(H)	H			
22 JUL (204) 1996	10:33	TPT	(A)	H			
23 JUL (205) 1996	03:12	RUV	(A)	H			
23 JUL (205) 1996	03:12	TIA	(H)	H			
23 JUL (205) 1996	03:12	TPT	(A)	H			
23 JUL (205) 1996	05:20	AFR	(H)	S	Kermadec	2.3×10^{25} dyn-cm	l
23 JUL (205) 1996	05:20	PPT	(H)	S			
23 JUL (205) 1996	13:24	PPT	(H)	H	Hawaii	4.6 m_b	g
23 JUL (205) 1996	13:24	RUV	(A)	H			
23 JUL (205) 1996	13:24	TIA	(H)	H			
23 JUL (205) 1996	13:24	TPT	(A)	H			
29 JUL (211) 1996	04:06	PMO	(A)	V	Hawaii	3.9 M_D	g
29 JUL (211) 1996	04:06	RUV	(A)	V			
29 JUL (211) 1996	04:06	TPT	(A)	V			
29 JUL (211) 1996	04:08	RUV	(A)	V	Hawaii		
29 JUL (211) 1996	04:08	TIA	(A)	V			
29 JUL (211) 1996	04:08	TPT	(A)	V			
29 JUL (211) 1996	11:00	PPT	(H)	H	Hawaii	4.4 M_D	g
29 JUL (211) 1996	11:00	RUV	(A)	H			
29 JUL (211) 1996	11:00	TPT	(A)	H			
14 NOV (319) 1996	13:47	PPT	(H)	S	S. of Fiji	1.9×10^{25} dyn-cm	m
16 MAR (075) 1997	14:19	PMO	(A)	H	Hawaii	4.2 M_D	
16 MAR (075) 1997	14:19	PPT	(H)	H			
16 MAR (075) 1997	14:19	TIA	(H)	H			
16 MAR (075) 1997	14:19	TPT	(A)	H			

Table 1

(Continued)

Date	Origin Time	Station		Nature	CODE NAME (E) or Region	Event size	Reference
D M (J) Y	GMT	Code	Site	(†)	(E, S, H, V)	(*)	(§)
16 MAR (075) 1997	14:19	TVO	(H)	H			
30 JUN (181) 1997	15:47	PMO	(A)	H	Hawaii	3.7×10^{24} dyn-cm	n
30 JUN (181) 1997	15:47	PPT	(H)	H			
30 JUN (181) 1997	15:47	RKT	(H)	H			
30 JUN (181) 1997	15:47	TBI	(H)	H			
30 JUN (181) 1997	15:47	TIA	(H)	H			
30 JUN (181) 1997	15:47	TPT	(A)	H			
30 JUN (181) 1997	15:47	TVO	(H)	H			
8 JUL (189) 1997	12:11	PPT	(H)	S	Aleutian	6.7×10^{24} dyn-cm	o

(†) E: Underwater explosion; H: Earthquake at hotspot volcanic site; M: Missile fired from submarine; N: High-altitude nuclear test; S: Subduction zone earthquake; V: Explosive volcanoseismic event.

(*) The size of an earthquake (event type S or H) is given when available by its seismic moment M_0 (in units of dyn-cm), otherwise by its PDE body-wave magnitude m_b, or if unavailable by its local magnitude M_L assigned by RSP, or its duration magnitude M_d, assigned by Hawaiian Volcano Observatory. When available, the size of an explosive (E) event is given by its yield expressed in tons of TNT (t).

(§) References to indidual events: a: TALANDIER and OKAL (1982); b: TALANDIER and OKAL (1984a); c: TALANDIER and OKAL (1987b); d: TALANDIER and OKAL (1984b); e: NAVA et al. (1988); f: WEIGEL et al. (1990); g: PDE (USGS); h: DZIEWONSKI et al. (1996); i: DZIEWONSKI et al. (1997a); j: DZIEWONSKI et al. (1997b); k: DZIEWONSKI et al. (1997c); l: DZIEWONSKI et al. (1997d); m: DZIEWONSKI et al. (1998); n: DZIEWONSKI et al. (1999a); o: DZIEWONSKI et al. (1999b).

the two types of structure, the low slope of high islands giving rise to a more complex conversion, and hence increasing the duration of the converted signal. In practice, we emphasize in the present study records obtained on atoll stations, where the conversion mechanism is simpler, and hence the records cleaner.

The sources of the *T* waves can be classified as either natural or man-made. Among natural sources, we distinguish between subduction zone earthquakes, intraplate earthquakes, and explosive underwater volcanic events. Among man-made sources, we identify chemical explosions in the ocean, and presumed firings of missiles from submarines. Among predigital events, we also consider four large atmospheric nuclear tests over Christmas Island in 1962.

Digital Processing

We base our search for satisfactory criteria of identification of the nature of *T*-wave sources on a systematic processing of the records along the following algorithm.

- RSP digital time series $s(t)$, available at a sampling rate $\delta t = 0.02$ s, are exemplified in the top frames of Figure 1. The length of the time windows analyzed is generally 20.46 s (1024 points) for the smaller signals, or 81.90 s (4096 points) for *T* waves emanating from larger subduction zone earthquakes. All

Table 2

T-wave records used in study of high-energy sources

Date	Time	Station		Nature	CODE NAME (N) or Region	Event size	Reference
D M (J) Y	GMT	Code (†)	Site	(*)	(E, P, S, H)	(**)	(§)
Analog (paper) records from explosive sources							
15 JUN (166) 1962	16:01	PPT	(H)	N	DOMINIC	0.8 Mt	a
30 JUN (181) 1962	15:21	PPT	(H)	N	DOMINIC	1.3 Mt	a
10 JUL (191) 1962	16:33	PPT	(H)	N	DOMINIC	1.0 Mt	a
11 JUL (192) 1962	15:37	PPT	(H)	N	DOMINIC	3.9 Mt	a
24 MAY (144) 1966	05:49	TAH	(H)	E	California	1 kt	b
06 SEP (250) 1968	02:07	RGI	(A)	E	Aleutian Is.	0.34 kt	b
13 AUG (225) 1969	16:12	RGI	(A)	P	Vancouver Is.	4.6 m_b	b
09 SEP (252) 1969	21:53	RGI	(A)	P	Vancouver Is.		c
01 OCT (274) 1969	17:11	RGI	(A)	P	Vancouver Is.	4.7 m_b	b
28 MAY (148) 1970	17:38	RGI	(A)	P	Vancouver Is.	4.9 m_b	b
04 SEP (247) 1970	21:23	RGI	(A)	P	Vancouver Is.		c
Additional earthquake records (digital)							
29 NOV (333) 1975	14:47	TPT	(A)	H	Kalapana	1.8×10^{27} dyn-cm	d
22 JUN (173) 1977	12:08	PAE	(H)	S	Tonga	1.8×10^{28} dyn-cm	e
15 NOV (319) 1994	23:10	RUV	(A)	S	Vancouver Is.	3.0 M_L	c

(†) TAH: Parameters at PMO extrapolated from unclipped hydrophone record at Papeete; RGI: Parameters estimated for clipped records at PMO from unclipped records across Rangiroa Atoll.

(*) N: Atmospheric nuclear test; E: Confirmed underwater explosion; P: Presumed underwater explosion; H: Earthquake at intraplate hotspot site; S: Subduction zone earthquake.

(**) The size of explosions (N, E) is given by their yield in tons of TNT, as announced by the appropriate agencies. For presumed underwater explosions (P), their size is given, when available, by the body-wave magnitude m_b reported by the USGS. The size of the large earthquakes (event type S or H) is given by their seismic moment M_0 (in units of dyn-cm). Note that the Kalapana event was accompanied by a large submarine slump (MA *et al.*, 1999), so that a moment tensor description may be inappropriate. For the small 1994 event, we give the local magnitude M_L computed by the Pacific Geoscience Centre (Sidney).

(§) References: a: ANONYMOUS (1989); b: *PDE* (USGS); c: This study; d: ANDO (1979); e: LUNDGREN and OKAL (1988).

signals are high-pass filtered ($f \geq 2$ Hz) to emphasize frequencies capable of propagating in the SOFAR channel.

• In order to eliminate rapid fluctuations in the waveshape, the *envelope e(t)* is then obtained by taking the absolute value of the signal, $|s(t)|$, and computing its 1-s (50-point) average in a moving window sliding in increments of 1 sample (0.02 s). The resulting series is then run through a 1-s moving window running average, to eliminate any remaining higher-frequency fluctuations. The purpose of this smoothing is to recover the general shape of the waveform, regardless of rapid fluctuations which may be due to local recording conditions and may not be representative of the true energy in the T wave. Examples of envelopes are shown in the central frames of Figure 1. The width of the sliding window (1 s) was chosen by trial and error between 0.5 and 2.5 s. The general amplitude of the record is characterized by the maximum value, e_{Max}, of $e(t)$.

- We then compute the *spectrum* of the signal, shown in the bottom frames of Figure 1. Because of limitations inherent in the response of the seismic sensors used at the RSP, we limit all our frequency investigations to $f \leq 16$ Hz.
- Finally, a *spectrogram* is obtained through a classical frequency-time analysis. For shorter signals, windows are 1.26 s (64 points) long, and are lagged 0.64 s (32 points), while for longer signals, the windows and lags are quadrupled. The frequencies studied are between 1 and 16 Hz. Examples of spectrograms are shown on Figure 2.

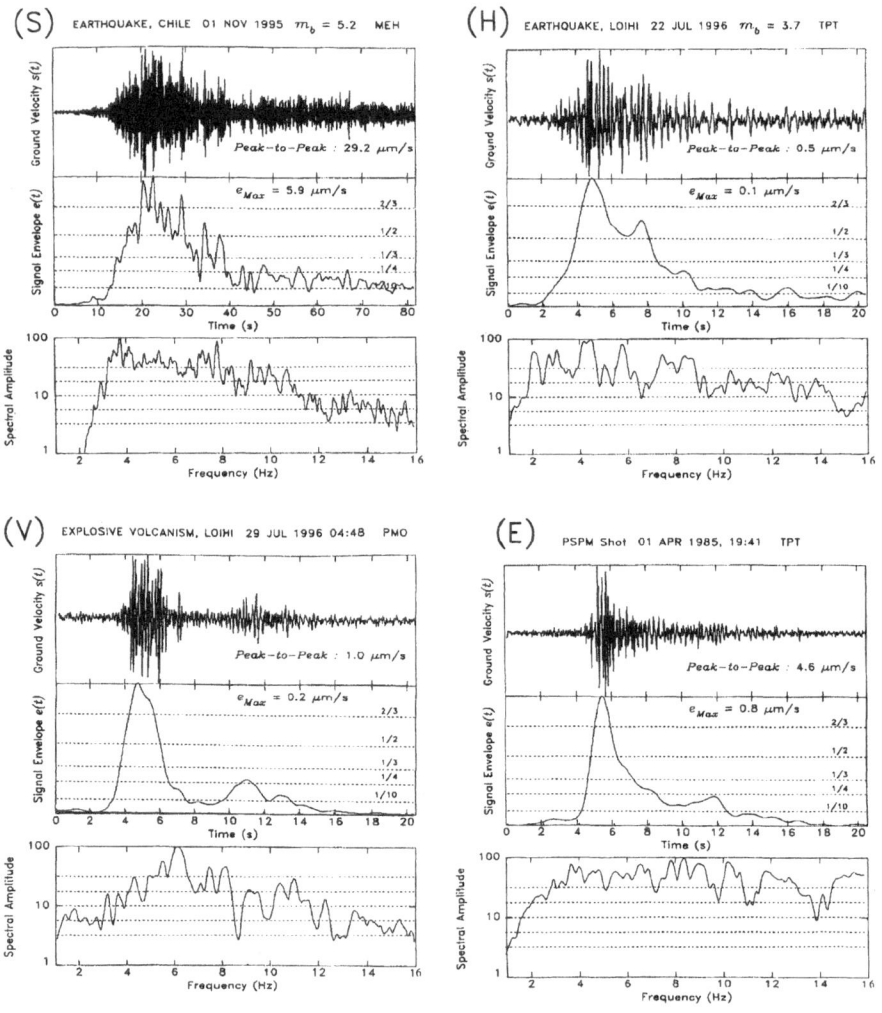

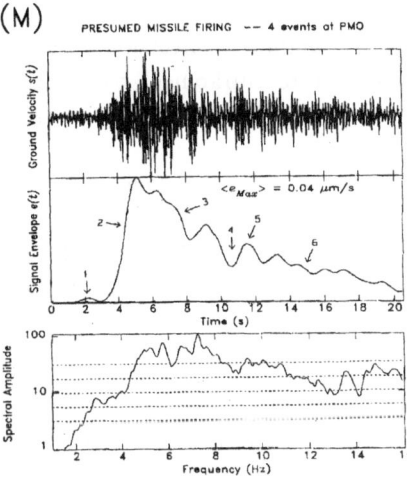

Figure 1

Typical examples of various records of T waves used in this study. **(S):** Subduction zone earthquake; **(H):** Intraplate earthquake; **(V):** Explosive event during a volcanic swarm; **(E):** Man-made underwater explosion; **(M):** Presumed firing of missile from submarine. In each case, the top frame is the deconvolved ground velocity time series $s(t)$ (note the longer window for the subduction event), the middle frame is the envelope $e(t)$, and the bottom frame the spectral amplitude of the signal $s(\omega)$, scaled logarithmically to its maximum value. The horizontal dashed lines in the center frames are the threshold used in the determination of the durations τ_r; similarly those in the bottom frames are the thresholds used in the determination of the width of the spectrum. Because of their low amplitudes, four missile signals have been stacked (with time lags obtained from maximizing their cross correlations) to produce Frame (M).

General Characteristics of Seismograms

Based on typical examples shown on Figures 1 and 2, we describe in this section a number of very general characteristics of the various kinds of sources used.

1. Subduction zone earthquakes (S)

We selected 30 records from 11 earthquakes from the years 1995–1997; a typical T wave is shown on Figure 1(S), and a spectrogram on Figure 2(S).

T-wave records from subduction events usually exhibit a combination of low frequencies, long durations, and emergent waveforms resulting in a spindle-shaped envelope. This can be grossly explained by the following combination of factors: First, the hypocenter of a subduction zone earthquake, typically located at the interface of the colliding plates, will be substantially removed from the water column. The resulting land path before conversion will be long, with high frequencies effectively attenuated before conversion to acoustic energy. The long land path on the source side also favors multipathing, leading to a prolonged waveshape. In addition, subduction events can be large, with sources extending in time for several tens of seconds, and contributing to the duration of the T-wavetrain

(OKAL and TALANDIER, 1986). Finally, the conversion slope will often be shallow-dipping, which requires a complex conversion mechanism involving several reverberations in the water before entrapment by the SOFAR channel can take place, hence the emergent nature of the wavetrain (TALANDIER and OKAL, 1998; PISERCHIA et al., 1998).

We also include in this category T waves from a few transform fault earthquakes in the South Pacific, which were found to exhibit duration characteristics similar to those of subduction events. This can be explained by the absence of steeply dipping converter slopes at the source, even though the hypocenters of the shallower transform earthquakes are located closer to the water mass than those of subduction events.

2. Intraplate earthquakes originating at hotspots (H)

We study 33 records from 10 events with epicenters at the Hawaiian and Society hotspots. A typical example is shown on Figure 1(H) and a spectrogram on Figure 2(H).

In contrast to subduction events, such earthquakes are generally of smaller magnitudes, and very shallow. As such, the shorter seismic paths on the source side allow retention of high frequencies, resulting in a generally whiter spectrum. Also, in geometries such as that of the South Coast of the Big Island of Hawaii, the existence of steep slopes at the head of individual basaltic flows allows a direct and efficient transfer of energy into the SOFAR channel, which results in a generally impulsive wavetrain (TALANDIER and OKAL, 1998).

3. Volcanoseismic explosions (V)

We use 17 records from 8 events, selected from representative periods of activity at Loihi (Hawaii), Macdonald Seamount (Southcentral Pacific), and Ruby (Mariana Islands). A typical signal is shown on Figure 1(V) and a spectrogram on Figure 2(V).

It has long been known that episodes of underwater volcanic activity feature explosive events often interpreted as the release of magmatic conduits opening the way for the eruption of lava into the ocean (TALANDIER and OKAL, 1987a). In particular, such events were observed systematically at the beginning of swarms at Macdonald Seamount (TALANDIER and OKAL, 1982). Since they are presumed to occur at the solid-water interface, they directly generate abundant T waves of relatively short duration, and in the absence of a source-side seismic path, they can feature a high-frequency spectrum. However, certain sources such as Macdonald have spectra limited to $f \leq 10$ Hz.

The explosive nature of these natural sources renders their correct identification and discrimination from man-made explosions a clear challenge. In this respect, it is important to note that volcanoseismic explosions most often

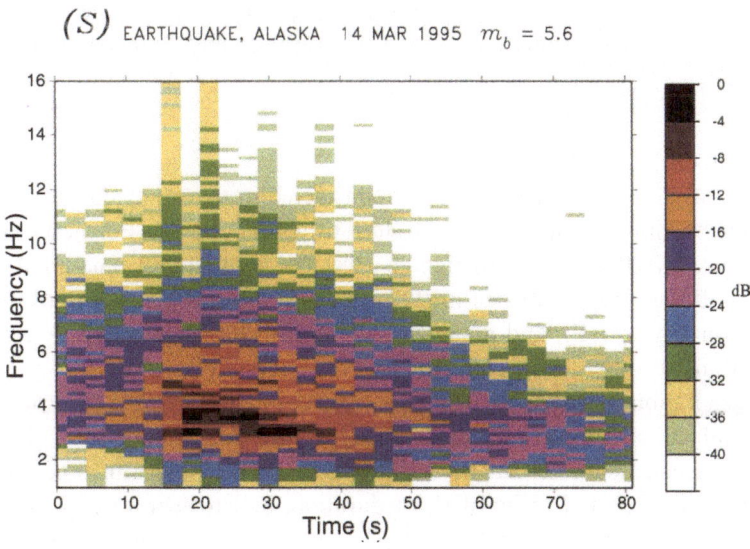

Figure 2(S)

Spectrogram of a typical record from a subduction zone event. The spectral amplitude of ground velocity $(s(\omega, t)$, in units of microns) present in each time–frequency pixel is coded according to the scale bar at right (in dB from maximum spectral amplitude, i.e., $20 \log_{10}[s(\omega, t)/s_{max}]$). Note the predominance of low frequencies.

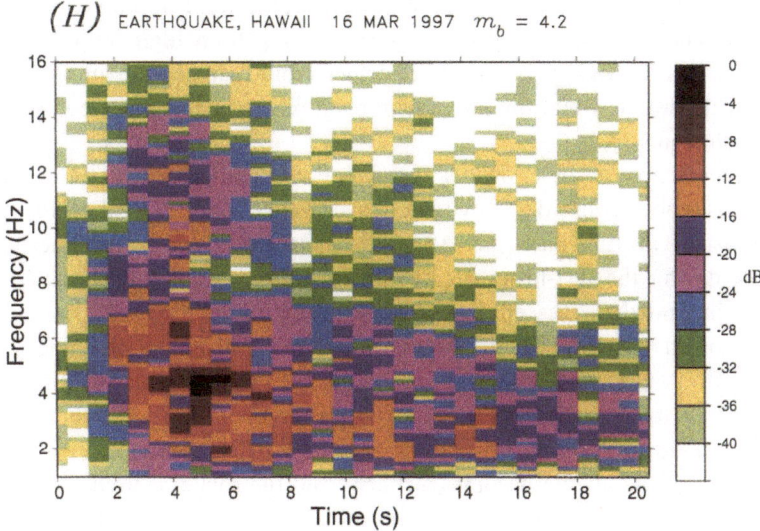

Figure 2(H)

Spectrogram of a typical record from an intraplate earthquake. Note presence of higher frequencies (10–12 Hz) in the early parts of the records.

take place in repetitive, albeit not periodic, sequences. This is clearly shown on Figure 2(V), where a second event is present approximately 12 seconds after the main one.

4. Chemical explosions (E)

The main group of explosive sources considered in this study consists of seismic refraction experiments conducted over close to two decades in Polynesia, including the extensive MIDPLATE campaign in 1989 (WEIGEL, 1990); and of the PSPM campaign off the coast of Mexico in 1985 (NAVA et al., 1988). An example of record is shown on Figure 1(E), and a spectrogram shown on Figure 2(E). A total of 64 records from 30 sources were studied.

The generation of sound by explosive sources in the oceanic environment has been the subject of considerable research (e.g., CHAPMAN, 1985). In the framework of this study, such sources can be considered as point sources in time and space. The signals are expected to be of short duration, and to feature high frequencies, but are made more complex by the bubble resonance at a frequency controlled by a combination of source yield and depth (COLE, 1948). However, some firing strategies used in seismic refraction campaigns aim at minimizing the bubble effect (WEIGEL, 1990).

5. Presumed missile firings from submarines (M)

During the mid-1980s, a number of T waves with very low amplitudes and singular characteristics were recorded principally at Station PMO on Rangiroa Atoll. These were characterized by spectrograms featuring both high frequencies, and a long-source duration (Fig. 2(M)). The former would suggest an impulsive source directly in the water, while the latter requires a phenomenon more complex than a simple explosion.

The occurrence of these signals correlated systematically with the issuance of marine advisories prohibiting navigation over vast expanses of the Northwestern Pacific Basin. On this basis, we speculate that the T waves in question were generated during the test-firing of missiles by submerged submarines. A possible scenario, illustrated on the envelope frame of Figure 1(M), would include: (1) a small explosion possibly due to the opening of a valve and the release of the missile; (2) a fast and sharp growth of the signal with white spectrum, corresponding to the underwater firing of the missile; (3) to (4) the slow decay of the signal during the underwater propagation of the missile; (5) a reburst of amplitude when the missile becomes airborne and the rocket is fired; and finally (6) a slow decrease in amplitude during the initial propagation of the rocket in the atmosphere.

Six such records are included in our database.

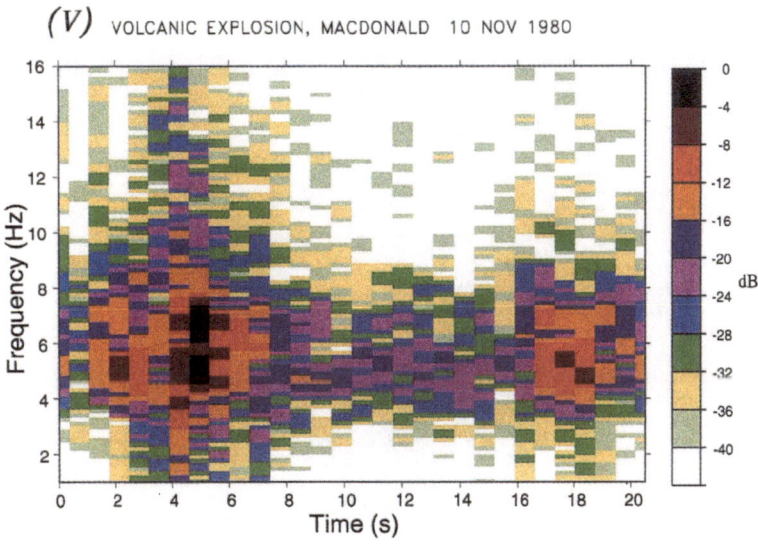

Figure 2(V)
Spectrogram of a typical record from an explosive event during a volcanoseismic swarm at Macdonald Volcano. Note the second event, 17 s into the record, illustrating the repetitive character of this kind of source.

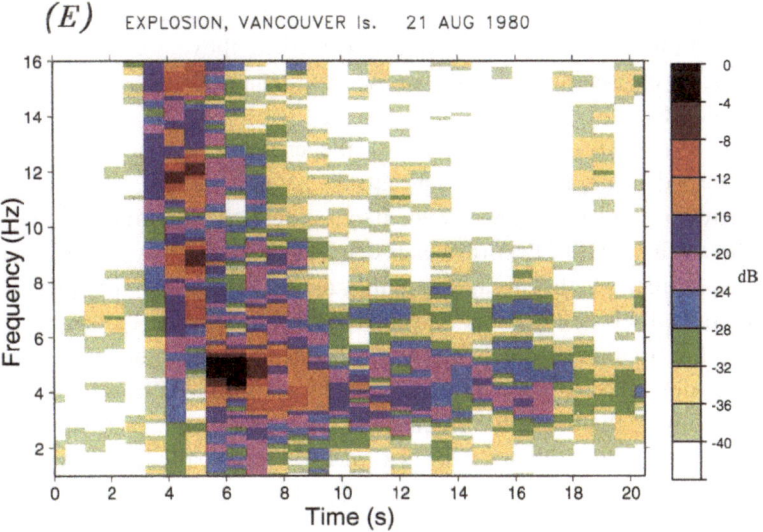

Figure 2(E)
Spectrogram of a typical record from a man-made underwater explosion.

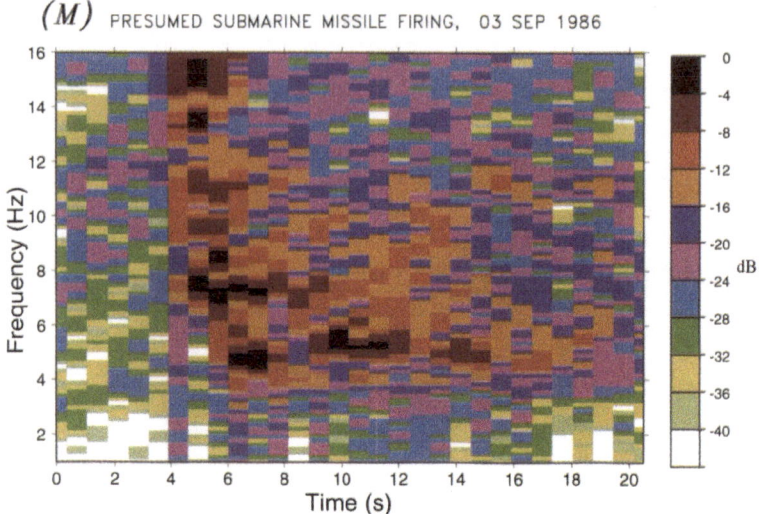

Figure 2(M)

Spectrogram from a presumed underwater missile firing. Note the scattering of the energy over much of the time–frequency plane.

3. Identification of Discrimination Criteria

In this section, we seek robust criteria allowing the discrimination between natural and man-made sources of T waves received at oceanic islands. For this purpose, and on the basis of the processing described above, we define the following parameters, as characteristics of the individual signals. We first focus on the time domain, and compute:

1. The *Maximum Amplitude* e_{Max} of the envelope of the signal. This is of course expected to depend on the epicentral distance traveled by the T wave. We choose here to correct our measurements to a reference distance of 3000 km (or 27° at the surface of the Earth), which can be regarded as typical of transpacific paths. The origin of the decay in amplitude of T waves with distance is, in principle, three-fold: (i) the effect of geometrical spreading at the spherical surface of the Earth; (ii) the effect of intrinsic dispersion during the propagation, and (iii) the possible effect of anelastic attenuation of sound waves in the water. The latter is usually negligible at the frequencies considered here. Because T waves can be considered as mildly dispersed surface waves, the global effect of (i) and (ii) is a decay of amplitude with angular distance Δ, varying like $1/\sqrt{\Delta \sin \Delta}$ (e.g., OKAL, 1989). Thus we will use, in all further discussion and figures, amplitude maxima *corrected* for distance according to:

$$e_{\text{Max}} = e_{\text{Max}}^{\text{Corrected}} = e_{\text{Max}}^{\text{Raw}} \cdot \sqrt{\frac{\Delta}{\Delta_{\text{Ref}}} \frac{\sin \Delta}{\sin \Delta_{\text{Ref}}}} \tag{1}$$

where $\Delta_{\text{Ref}} = 27°$. A number of other empirical distance corrections were tried, notably power laws of the form Δ^{α}, with α ranging from 0.5 to 1.5, with no appreciable influence on our results. This technique obviously requires that the distance to the source be known. In practice, this can be achieved with a regional network, such as the RSP, using at least three stations.

2. The *Duration* τ_r of the signal, defined from the number of points in its envelope reaching a given threshold, $r = 1/10, 1/4, 1/3, 1/2$ or $2/3$, of the maximum amplitude e_{Max}; we will focus primarily on the duration at $1/3$ of maximum, $\tau_{1/3}$.

A difficulty arises from the presence of background noise in the records, the latter varying significantly, not only as a function of time, but even among different stations recording the same event. In order to alleviate this problem, we define the noise level n by considering a 2–second window of record before the arrival of the T phase, and computing the maximum value of its envelope by the same algorithm as in 1. above. The duration τ_r is then defined from the number of points for which the envelope exceeds the level $(n + r \, e_{\text{Max}})$.

3. The *Rise Time* T_R, defined as the time it takes the envelope of the signal to grow from 25% to 100% of its maximum amplitude.

4. Similarly, the *Fall Time*, T_F, defined as the time it takes the envelope of the signal to fall from 100% to 25% of its maximum amplitude.

5. The *Duration of maximum intensity of the signal*, defined as the sum $T_R + T_F$, and representing the time during which the signal keeps a sustained amplitude above $r = 1/4$ of its maximum.

It can differ from the duration $\tau_{1/4}$ as defined in (2), since a complex signal may occasionally fall below the threshold value, and then cross it again. This is especially true for many (but not all) volcanic sources, some intraplate earthquakes, and the largest subduction events.

6. The integrals $I_{\text{env}} = \int e(t) \, dt$ and $J_{\text{inv}} = \int e^2(t) \, dt$ of the envelope over the full window analyzed.

7. The *Skewness Sk* and *Kurtosis Ku* of the signals are defined from their envelope time series $e(t)$ as:

$$Sk = \frac{\int_0^T e(t)(t - \bar{t})^3 dt / \int_0^T e(t) \, dt}{\sigma^3} \tag{2}$$

and

$$Ku = \frac{\int_0^T e(t)(t - \bar{t})^4 dt / \int_0^T e(t) \, dt}{\sigma^4} - 3 \ . \tag{3}$$

These quantities are moment estimators which describe the general shape of a statistical variable; in these formulae, $\bar{t}$ and σ are the centroid time of the envelope, and the standard deviation of t, respectively:

$$\bar{t} = \frac{\int_0^T e(t)t\,dt}{\int_0^T e(t)\,dt}; \qquad \sigma^2 = \frac{\int_0^T e(t)(t-\bar{t})^2\,dt}{\int_0^T e(t)\,dt}. \tag{4}$$

Because they feature a slower buildup of strain release, as well as a longer source duration, earthquake sources would, in principle, be expected to generate signals with lower values of both Sk and Ku, as compared to higher values of these coefficients for explosion signals.

8. In addition, we also investigate the correlation between the shape of a signal's envelope, and that of reference signals, as detailed in Section 3.

9. In the frequency domain, we define:

The *mean frequency*, $\langle f \rangle$, as the average of those frequencies maximizing the spectral amplitude in each sliding time window of the spectrograms;

10. The *maximum frequency*, f_{Max}, as the greatest among those frequencies maximizing the spectral amplitude in each sliding time window of the spectrograms;

11. The *width of the frequency spectrum*, measured as the number of points, in the spectral domain, for which the spectral amplitude $X(\omega)$ is greater than a given fraction a of its maximum value. This parameter is computed for the values $a = 1/3.2,\ 1/5.6,\ 1/10,\ 1/32$ and $1/56$, which constitute a geometrical progression of ratio $r = 10^{-1/4}$.

12. Finally, we consider the *ratio of low- and high-frequency energy* (R_{LH}), defined as the ratio of the integral of the energy in the two bandwidths $2 \leq f \leq 9$ Hz and $9 \leq f \leq 16$ Hz.

The Failure of Simple Criteria in the Frequency Domain

The idea of seeking a discriminant in the frequency characteristics of the signals stems from at least two properties: first, explosions are expected to be shorter-lived than earthquakes, thus generating high-frequency spectra; and second, earthquake sources being removed from the conversion point, will see their higher frequencies attenuated over the source-to-conversion path. One would then expect explosions to be separable from earthquakes, on the basis of a higher-frequency spectrum.

However, the examination of the spectra on Figures 1(H) and 1(E) shows that both are remarkably "white." Similarly, the spectrograms on Figures 2(H) and 2(E) show that both kinds of signals can feature high frequencies (in the range of 10–12 Hz) in the earliest parts of the signal. In order to analyze this situation further, we carried out the following investigations:

1. Mean frequency

On Figure 3a, we plot the maximum envelope amplitude e_{Max} against $\langle f \rangle$. While the lowest mean frequencies ($\langle f \rangle \leq 4$ Hz) are exclusively found in records of earthquakes (subduction or intraplate), and the highest ones ($\langle f \rangle \geq 7.5$ Hz) for explosions (man-made or volcanic), the intermediate field of $\langle f \rangle$ values is populated with all types of records. No trend with amplitude is present, and no separation can be achieved, especially between man-made explosions and intraplate earthquakes.

2. Maximum frequency

As shown on Figure 3b, the results are fundamentally similar (except of course for a change of frequency scale) when using f_{Max} instead of $\langle f \rangle$.

3. Width of spectrum

Figure 4 plots e_{Max} against the width of the spectrum, measured at 1/5.6 of its maximum value. Once again, it is clear that no separation between the various kinds of sources can be achieved on the basis of these parameters. While it is generally true that subduction earthquakes have narrower spectra than other sources, intraplate earthquakes, especially those of smaller magnitude, can feature spectral widths as high as 13 Hz. Also, and rather surprisingly, we find that the spectral width of explosion signals vary over the whole range from 4 to 14 Hz. The elimination (in Fig. 4b) of stations on high islands does not improve the picture. Finally, a variation in the threshold used for the amplitude of the spectrum (from 1/3.2 to 1/56 of its maximum value) does not help either.

4. High-to-low frequency spectral ratio

Figure 5 explores the behavior of the ratio R_{LH} between the energy present in the signal in the low- and high-frequency bands. It is found to be largest ($R_{LH} > 10$) only for subduction earthquakes and certain volcanic explosions, and smallest ($R_{LH} < 2$) for man-made explosions and certain other volcanic events. However, the majority of the data points fall within the range $2 < R_{LH} < 10$ where the different types of sources cannot be separated. No correlation with source size (as expressed by e_{Max}) can be recognized. These results are not changed when stations from high islands are removed (Fig. 5b).

As a conclusion of these four tests in the frequency domain, we confirm that trends do exist along expected properties: subduction earthquakes have signals with generally lower frequencies and narrower spectra than do explosions. However, these trends are far from universal, and the large number of exceptions that they suffer prevents the use of any of the four studied criteria as a reliable discriminant between the various sources. In particular, intraplate earthquakes and explosions can have very similar frequency characteristics.

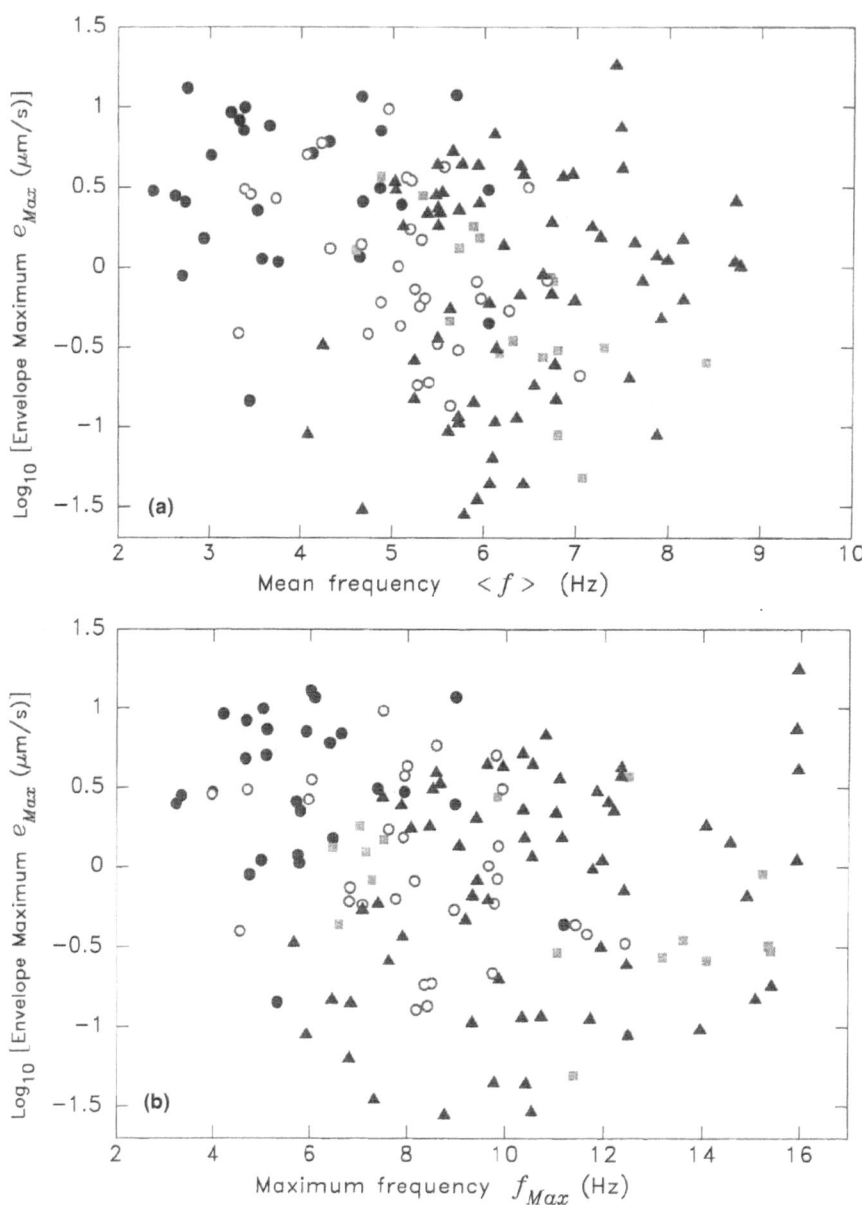

Figure 3

(a) Maximum of signal envelope, e_{Max}, plotted against the mean frequency $\langle f \rangle$. The following symbols are used to differentiate the various sources: *Solid circles*: Subduction earthquakes; *Open Circles*: Midplate earthquakes; *Solid squares*: Explosive volcanic events; *Triangles*: Man-made explosions. Records from both atoll and high island sites are used. Note that this property cannot be used to separate the various populations of events. (b): Same as (a) for the maximum frequency f_{Max}.

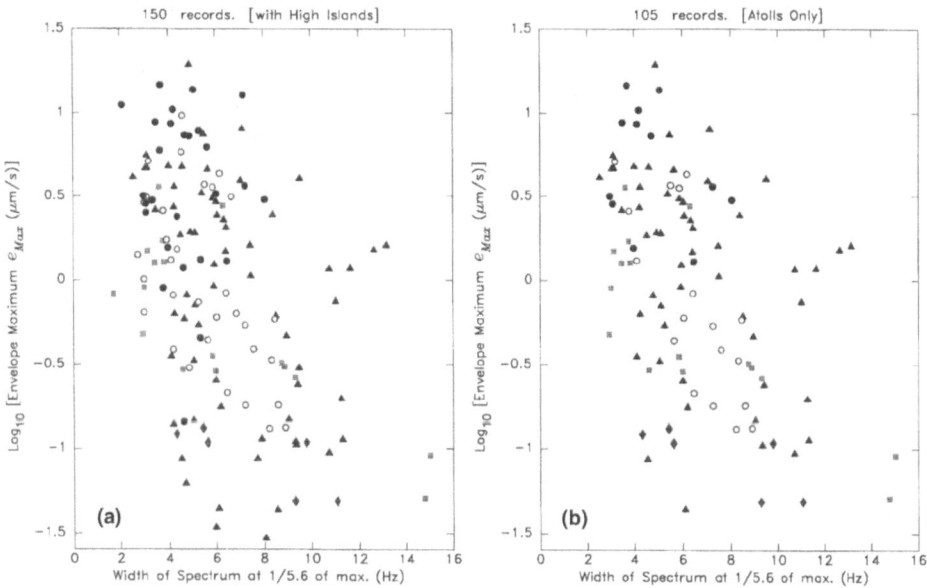

Figure 4

Maximum of signal envelope, e_{Max}, plotted against the width of the Fourier spectrum of the signal, measured at a fraction $a = 1/5.6$ of its maximum. (a): Entired data set; (b): Atoll stations only. Symbols as on Figure 3, complemented by *Diamonds*: Presumed firing of missiles. Note the absence of a clear separator.

Identification of Criteria in the Time Domain

We now turn our attention to potential discriminants related to time-domain characteristics of the signals.

1. Duration as a function of amplitude: The preferred discriminant

We compare here the maximum amplitude of the envelope of the T phase, e_{Max}, with its duration $\tau_{1/3}$ at $1/3$ of maximum (Fig. 6). It is clear from this figure that the comparison of these two parameters allows a general separation of earthquakes from explosions, with the former exhibiting a longer duration for a similar level of amplitude. We define empirically the line

$$\log_{10} e_{Max} = 4.9 \log_{10} \tau_{1/3} - 4.1 \tag{5}$$

where e_{Max} is in μm/s and $\tau_{1/3}$ in s, as an adequate separator in the $\tau_{1/3} - e_{Max}$ plane. We note that all subduction earthquakes fall clearly to the right of the line, as do all, except two, midplate earthquakes. These two exceptions are shown as bull's eye symbols on Figure 6. The first one, marginally misidentified by the separator, is a small earthquake at Mehetia ($M_L = 3.3$) recorded at VAH on the Southern Coast of Rangiroa Atoll. The second one concerns an earthquake in

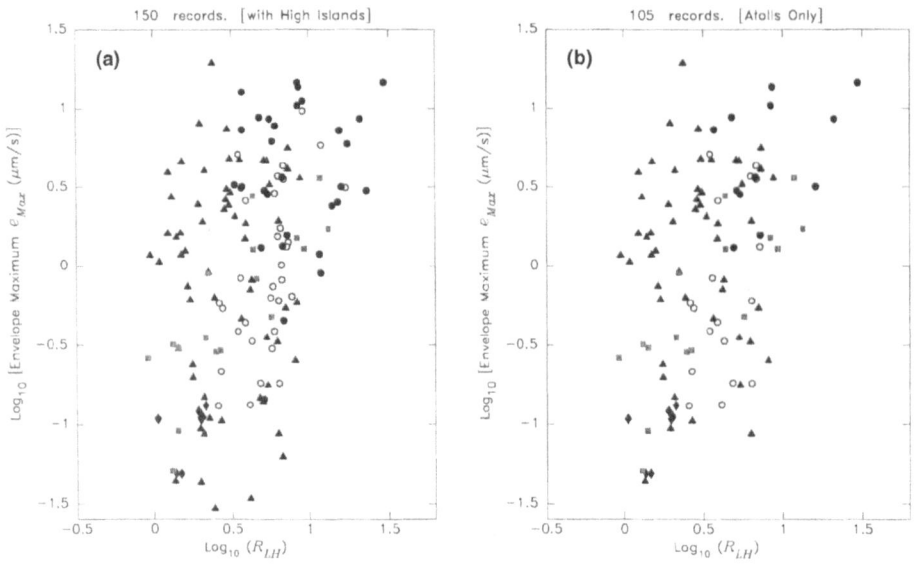

Figure 5

Maximum of signal envelope, e_{Max}, plotted against the ratio R_{LH} of the energies present at low and high frequencies. Symbols as in Figure 4. (a): Full data set; (b): Atoll sites only.

Hawaii ($m_b = 4.2$) recorded at TPT, on the Northern Coast of Rangiroa. It is worth noticing that the same event is correctly identified on its record at PMO. On the other hand, all man-made explosions fall to the left of the line. Volcanic explosive events cannot be discriminated. Finally, presumed firings of missiles (M) are distinguished from explosions, due to the length and complexity of the source process.

In a series of other trials, we investigated the sensitivity of the discriminant to the factor r (taken above as $1/3$) defining the duration τ_r. Similar attempts were performed with $r = 1/10$, $1/4$, $1/2$ and $2/3$, and the value $\tau = 1/3$ was found to provide the best results, between values of r too close to 1, for which the concept of duration loses its meaning, and small values which increase the window studied and allow possible contamination by low-energy scatterers.

Finally, on Figure 7, we explore the possibility of including records obtained at high island sites ("(H)" in Table 1). For the bigger events, with the larger envelope amplitudes, the data sets of explosions and earthquakes remain distinct, and a separator can be found (indeed the discriminant (5) remains effective). At the lower amplitudes, post-conversion scattering and multi-pathing on the receiver side can contribute substantially to the duration of the signal, even in the case of an explosion, and the separation becomes impossible. We verified that a similar pattern of degradation of the discriminant at high-island sites occurs for the other threshold levels r defining τ_r.

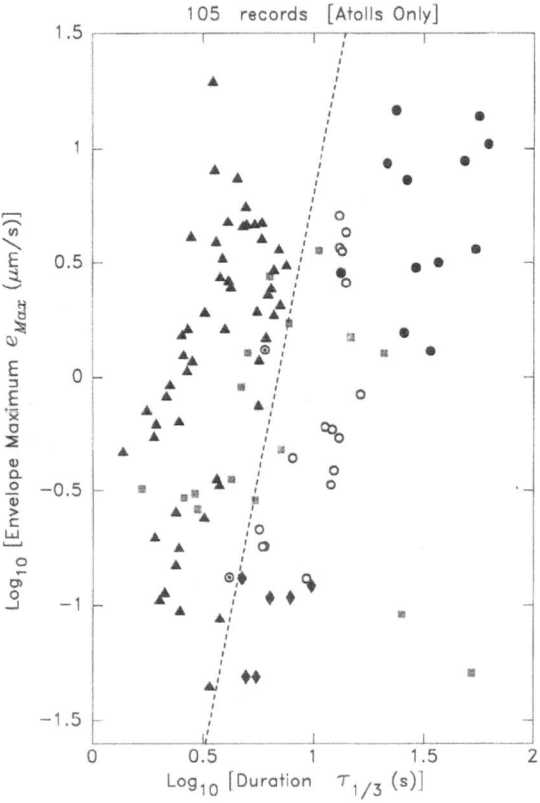

Figure 6

The duration-amplitude discriminant. This figure plots the envelope maximum e_{Max} as a function of signal duration measured at 1/3 of maximum, $\tau_{1/3}$. Symbols as in Figure 4, with the two bull's eye symbols showing the misidentified midplate earthquakes (see text for details). The dashed line is the separator proposed in Equation (5). This figure includes only records obtained at atoll stations.

The comparison between Figures 6 and 7 illustrates the importance of siting T–phase stations at shorelines with favorable conversion characteristics, i.e., on atolls or on volcanic structures featuring a steep slope (TALANDIER and OKAL, 1998).

2. Examination of other potential discriminants: Rise Time

Figure 8a explores the potential use of the rise time T_R as a discriminant. This is a legitimate suggestion, since the impulsive nature of an explosive source should lead to rise times shorter than for earthquakes. While subduction zone earthquakes are efficiently separated from explosions, a number of intraplate "hotspot" events are not. In seeking a possible explanation to this pattern, we note that the rise time of a T-wave signal is controlled by a combination of the rise time of the source (which can

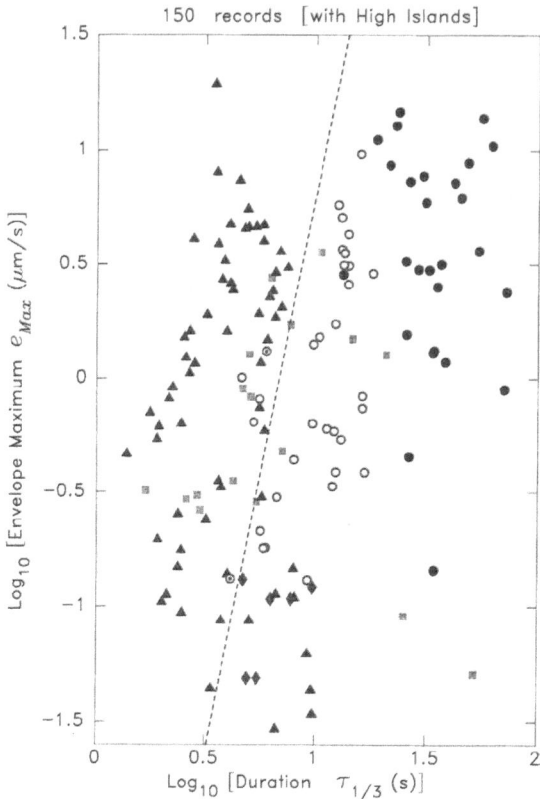

Figure 7
Same as Figure 6, but including records obtained at high island sites. Note the deterioration of the discriminant's performance.

be very short for small earthquakes), and of the buildup of the T phase at the conversion point. We have shown (TALANDIER and OKAL, 1998) that in favorable environments such as the southern shore of the Big Island, the T wave can be generated with a very impulsive wave shape. In addition, and especially for small events, the measure of rise time involves sub-second characteristics of the waveshape, which can then be affected by the slight dispersion observed over the longer paths. This effect is particularly visible at Rangiroa in the case of the 1985 PSPM shots.

In conclusion, we reject rise time as a satisfactory discriminant.

3. Fall time

We similarly investigate on Figure 8b the behavior of the fall time T_F. The situation is, if anything worse, in that even several subduction earthquakes now fail to be separated from the cluster of explosion records. We similarly discard fall time as a satisfactory discriminant.

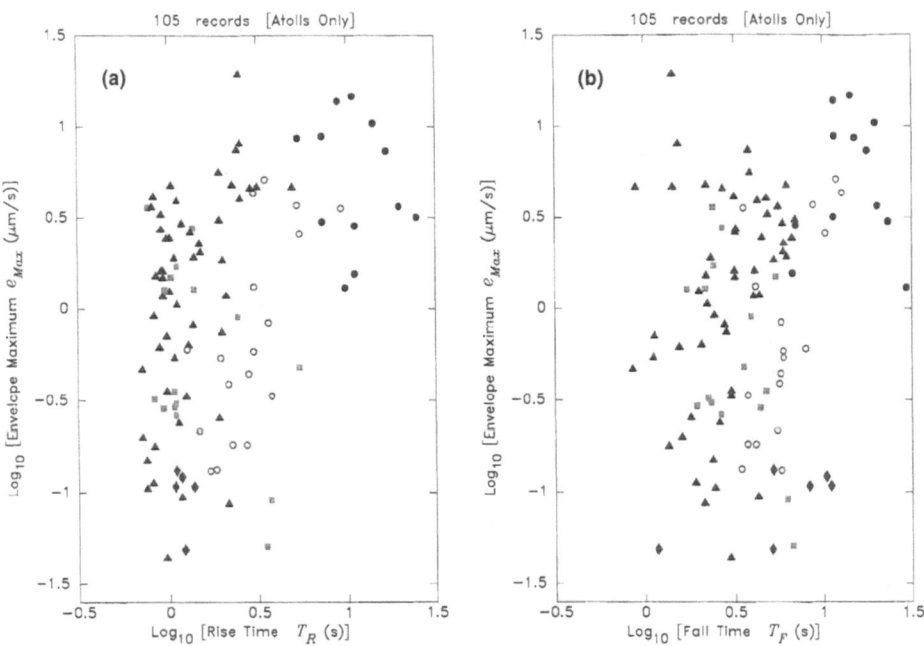

Figure 8

(a): Maximum envelope amplitude as a function of rise time T_R (symbols as in Fig. 4). Note that while subduction earthquakes are adequately separated from explosions by this criterion, intraplate events are not. (b): Same as (a) for the fall time, T_F. Note the failure of this parameter as an adequate discriminant.

4. Total duration T_{Total}

This parameter is the sum of the rise and fall times, and approaches the duration $\tau_{1/4}$ at the threshold used to define T_R and T_F. As expected, when plotted against the maximum amplitude derived from the envelope (Fig. 9a), this parameter does provide a reasonable discriminant. However, its performance is inferior to that of the duration $\tau_{1/4}$, and a fortiori, $\tau_{1/3}$, in particular for most intraplate earthquakes, and for the smaller explosions.

Another potentially interesting comparison is that of T_{Total} with the duration $\tau_{1/4}$ itself (Fig. 9b), which illustrates the smoothness of the envelope series $e(t)$. An explosion could be expected to have a simple source time history, and thus a smooth envelope and a T_{Total} identical to $\tau_{1/4}$; on the other hand, an earthquake's more complex source could lead to several crossings of the envelope threshold $r = 1/4$, and thus to a a a longer $\tau_{1/4}$, i.e., to a point plotting substantially below the dashed bisector on Figure 9b. Unfortunately, while this trend is indeed present, it is not universal: we find both earthquake records with smooth envelopes, even among the larger subduction events, and explosion records where T_{Total} is deficient by as much as a factor 2.5. In this respect, a discriminant based on this comparison would be far from foolproof, and cannot be proposed.

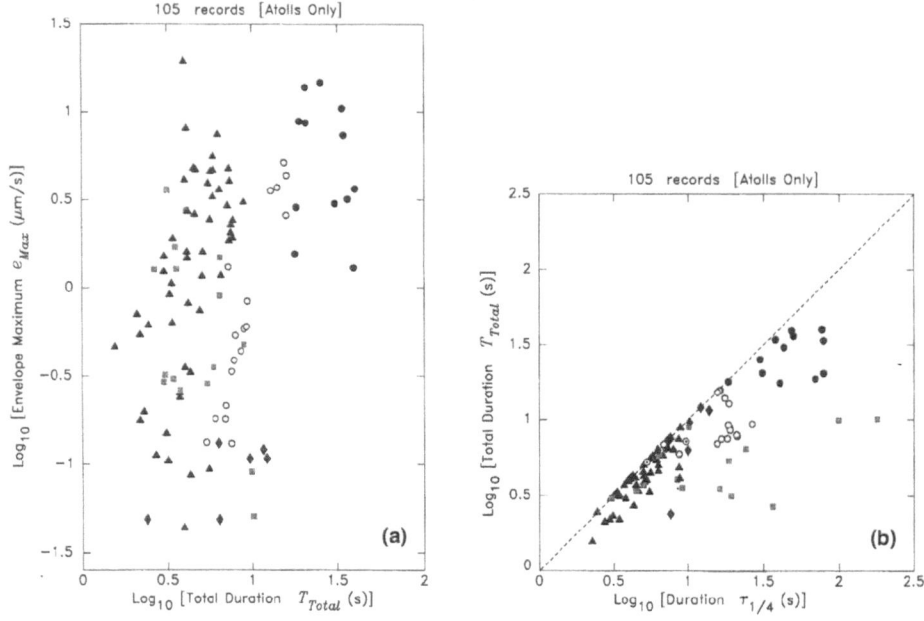

Figure 9

(a): Maximum envelope amplitude as a function of the total duration T_{Total} of sustained amplitude. Note the degradation of the separation in the case of smaller explosions. (b): Comparison of T_{Total} and $\tau_{1/4}$.

5. Skewness and Kurtosis

Pursuing a similar philosophy, we explore on Figure 10 the performance of the parameters Sk and Ku defined in Equations (2) and (3). The parameters are conveniently plotted against e_{Max} on frames (a) and (b), and against each other on frame (c). We find, once again, that the trend towards lower values of Sk and Ku for earthquake sources is correctly upheld by the end members of the distribution, but that the intermediate field ($Sk \sim 1$; $Ku \sim 0$) is populated by both earthquakes (mostly intraplate) and explosions. No trend with amplitude can be detected, and the inclusion of records at high island sites (d) worsens the situation. We thus discard skewness and kurtosis as effective discriminants.

6. Envelope integral

Figure 11 correlates the maximum envelope amplitude, e_{Max}, with the integral of the envelope, I_{env}, over the full time window retained for study. While the dashed line

$$\log_{10} e_{Max} = 9.20 \log_{10} I_{env} - 22.90 \tag{6}$$

does provide a separation between earthquakes and explosions, the latter is less frank and hence less adequate than the [$e_{Max} : \tau_{1/3}$] criterion described above in Equation

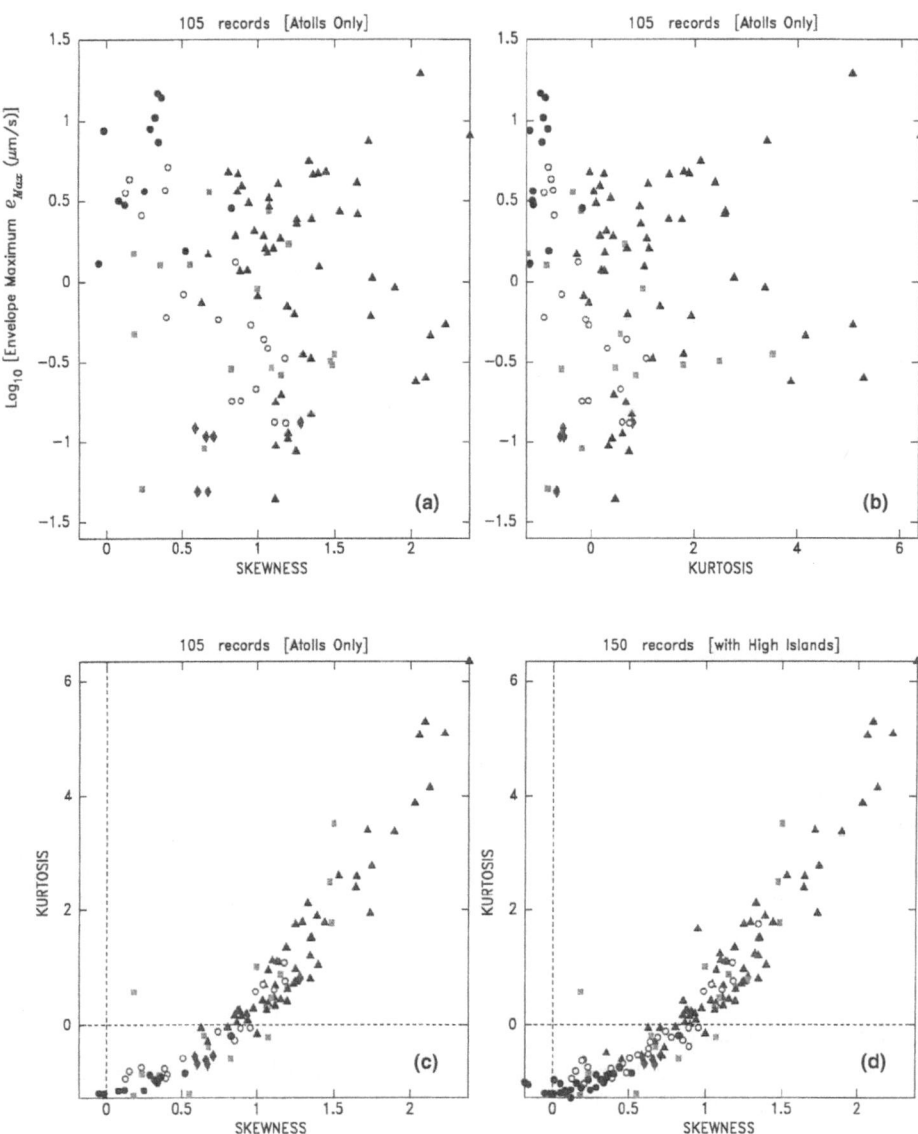

Figure 10
Performance of Skewness (a) and Kurtosis (b) estimators, as defined by Equations (2) and (3), and plotted as a function of e_{Max}, for 105 records at atoll sites. Frame (c) plots *Ku vs. Sk*, and (d) shows further deterioration of their performance when high-island sites are included.

(5). In particular, data points from the intraplate earthquakes are now regrouped very close to the separation, with three (as opposed to two on Figure 6a) straddling it. The use of the energy integral J_{env} instead of I_{env} would, if anything, degrade the results.

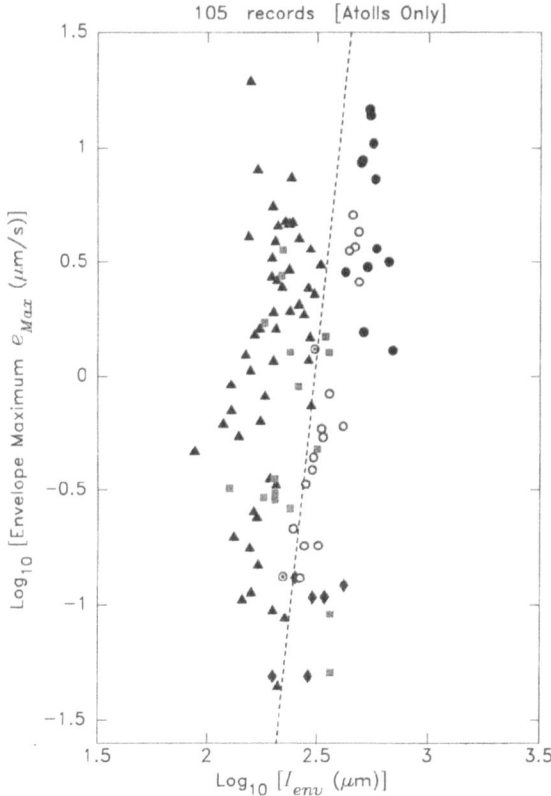

Figure 11

Maximum envelope amplitude as a function of the envelope integral I_{env}. Note that although the dashed line (Equation (6)) can be proposed as a discriminant, the separation is poorer than on Figure 6.

7. Reference envelopes

We define on Figure 12 *reference envelopes*, obtained by stacking, within each family of records (S, H, V, E), the envelopes of the relevant signals, after time-lagging them to align their maxima. The data set has been separated into nearby and distant events in order to examine the case of intraplate earthquakes (H) recorded at short distances. We confirm from this figure that the average waveshape of man-made explosion signals is similar to that of volcanic explosive signals, and to that of intraplate earthquakes, when the latter are recorded at short distance. On the other hand, distant earthquakes (of both types) have a characteristically broader envelope.

We further investigated these properties by computing for each record an envelope misfit, m, defined by taking the integral of the absolute value of the

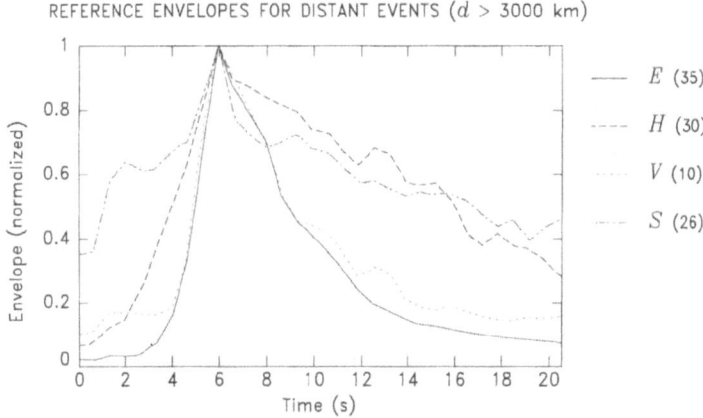

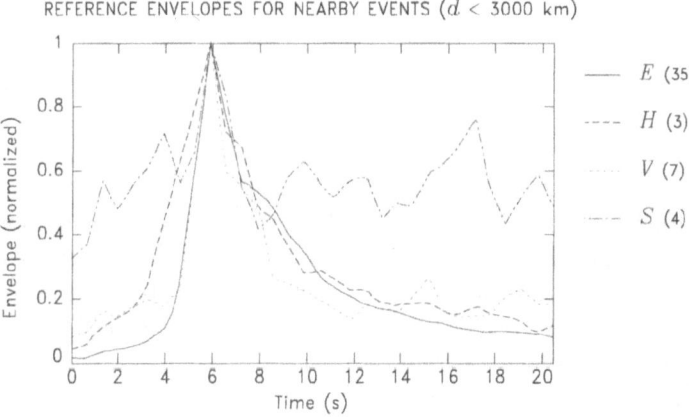

Figure 12
Reference envelopes computed by stacking the envelope signals $e(t)$ within each family of records. Given at
right is the number of records stacked to obtain each trace.

misfit between the signal's normalized envelope, and the reference envelope
for a particular class of signals, lagged by the time t' for which the maxima
coincide,

$$m = \int \left| \frac{e(t)}{e_{\text{Max}}} - e_{\text{Ref}}(t - t') \right| \cdot dt \; , \tag{7}$$

the domain of integration being that over which the envelope is at least 10% of its
maximum value, in order to avoid the contribution of background noise. Figure 13,
drawn in the case of reference to a subduction event envelope (all distances included),
shows that many records are indeed identified properly: subduction records have in

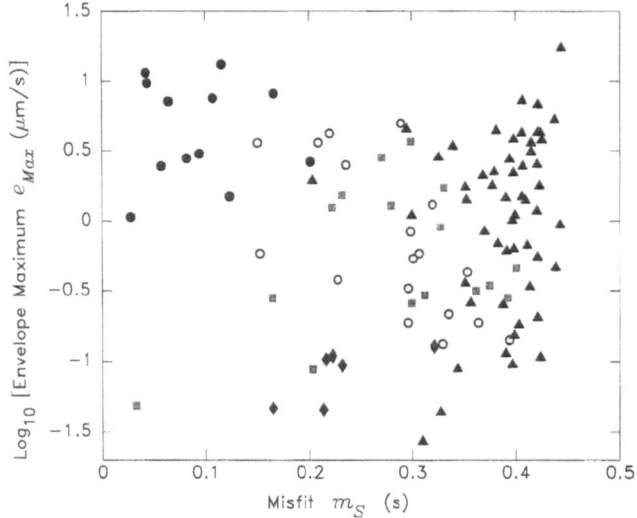

Figure 13
Misfit m_S computed relative to the subduction reference envelope.

general the lowest values of misfit m_S, and the largest ones are found for explosions. However, no reliable criterion for discrimination can be built, since the distinction between the various populations is not sharp, especially between explosions and hotspot earthquakes. We reached a similar conclusion with misfits m_E computed using the explosion envelope as reference.

4. Discussion

From the large number of tests described above, it appears that the comparison of the amplitude of the envelope of the signal, e_{Max}, and its duration at 1/3 of maximum, $\tau_{1/3}$, can be used as an efficient discriminant between tectonic earthquakes and underwater explosions. Another possible discriminant based on the envelope of the signal would use its maximum amplitude e_{Max} and its integral I_{env}. The former discriminant is preferred on account of its better resolution. However, in addition to the two small intraplate earthquakes mentioned above, the discriminant fails in a number of cases which deserve comment.

• Records from explosive events occurring during volcanoseismic swarms are not discriminated. Their characteristics can make them look either like earthquakes or explosions. Indeed, on all diagrams from Figures 3 to 13, the "V" records (plotted as squares) are found in the same fields as explosions (triangles) or earthquakes (especially intraplate; circles). To a large extent, this is due to the explosive nature of the phenomenon itself, which makes it futile to attempt to distinguish such sources from man-made explosions. However, one property can be used to discriminate easily between "V" and "E" sources: Volcanoseismic swarms are always long-lived on the scale of the duration of individual T-wave signals, with explosive events always repetitive, and their occurrence *not* periodic, but rather random in time. The amplitude of the signal also varies with each individual source. In addition, the individual events are separated by intervals of time featuring sustained T-wave activity at lower amplitudes, which TALANDIER and OKAL (1987a) have described as expressing the actual delivery of lava onto the ocean floor, following the opening of conduits during the explosive events. These properties are in contrast with our observations during occasional campaigns featuring multiple man-made explosions. As shown for example on Figure 1 of OKAL and TALANDIER (1986), the latter are characterized by perfect periodicity, both in time and amplitude. Also, no detectable activity is present between the various sources, when the T-wave signal falls back to the amplitude of background noise. This property makes it easy to discriminate volcanic explosions from man-made ones.

    • A second type of signals which does not lend itself well to discrimination is the set of presumed missile firings, although they are generally categorized as "earthquakes" under the discriminant (5), because of the complex nature of their source. Still, they can be recognized through their combination of very small amplitudes, long durations, and abundance of high frequencies. One data point (for the record of 19 August 1986) falls almost exactly on the separator on Figure 6. It features a significantly shorter signal than the other presumed missile firings, in particular those on the previous day; it also features greater values of the coefficients *Sk* and *Ku*. We speculate that it could represent an unsuccessful test, when the missile went astray and failed to become airborne.

Application to Larger Sources

    In the framework of the CTBT, it is also important to evaluate the performance of a discriminant such as $[e_{Max} : \tau_{1/3}]$ in the case of more energetic sources. Unfortunately, no man-made underwater explosions of truly large amplitude were recorded by RSP since the implementation of digital recording in the early 1970s. For this reason, we carry out in this section a more tentative analysis of analog (paper) records obtained in Polynesia from a number of very large explosions carried out in the 1960s and early 1970s and listed in Table 2. These include two major, announced, shots off California in 1966 (1 kt) and the Aleutians in 1968 (0.3 kt), and five events off the coast of Vancouver Island in 1969–1970. Among the latter, three were located by the USGS and ISC, within a few km of 48.47°N, 126.55°W, and given magnitudes m_b between 4.6 and 4.9. We located the other two at a similar epicenter from their T-phase records at the RSP. The nature of all five sources is unpublished, but based on the characteristics of their T waves, we propose that they are indeed explosions.

We also include in the data set digital records from three additional earthquakes. Two of them are events having generated exceptionally intense T waves, the 1977 Tonga event whose T waves woke up residents on the western shore of Tahiti (TALANDIER and OKAL, 1979) and the 1975 Kalapana earthquake, the largest recorded from a Pacific hotspot. The final event is a small ($M_L = 3.0$) 1994 earthquake located in the immediate vicinity of the Vancouver events.

Finally, we study paper records of T waves obtained at Papeete from four atmospheric nuclear tests carried out at Christmas Island in June-July 1962 as part of operation DOMINIC (ANONYMOUS, 1989).

Because no digital records are available for these large underwater events, no data processing can be performed in the frequency domain. In the time domain, the situation is aggravated by the fact that several records are significantly clipped. The 1966 Californian shot predates the instrumentation of Rangiroa Atoll, but we were able to extrapolate the amplitude of the unclipped record on a pressure sensor located in the Tahiti lagoon, to estimate an amplitude of 200 μm/s at station PMO (or 400 μm/s once corrected for distance). In the case of the more recent Aleutian and Vancouver shots, we similarly extrapolate the clipped amplitudes at TPT from unclipped records at other, less favorably located stations on Rangiroa Atoll, based on comparison with digital records of lower amplitudes obtained at later dates from similar locations. The resulting amplitudes must be considered tentative, and we assign them an error of a multiplicative or divisive factor of 2 (±0.3 logarithmic units). On the other hand, the duration of the signal can be measured from the unclipped records, with a precision of ~25%, or ±0.1 logarithmic unit. We show the resulting seven data points (with error ellipses) as the large triangles on Figure 14 (we use downward-pointing triangles for the five Vancouver events since their nature as explosions is unconfirmed).

It is remarkable that these seven data points and the two large Tonga and Kalapana earthquakes are perfectly separated by the $[e_{Max} : \tau_{1/3}]$ discriminant, which was derived for records with amplitudes at least one order of magnitude smaller. Based on the reported explosive nature of the California and Aleutian events, we conclude that the discriminant can be extended to the relevant domain of amplitudes, and we propose that the five Vancouver events are indeed man-made explosions. This interpretation is upheld by the qualitative observation of very high frequencies (estimated at 10 Hz) in the initial phases of the clipped records at TPT. In addition, the small 1994 earthquake selected for study with an epicenter only 40 km from those of the 1969–1970 events clearly has properties characteristic of a plate boundary ("S") event, and fundamentally different from those of the five 1969–1970 events. This effectively rules out the possibility that the latter could be seismic in origin.

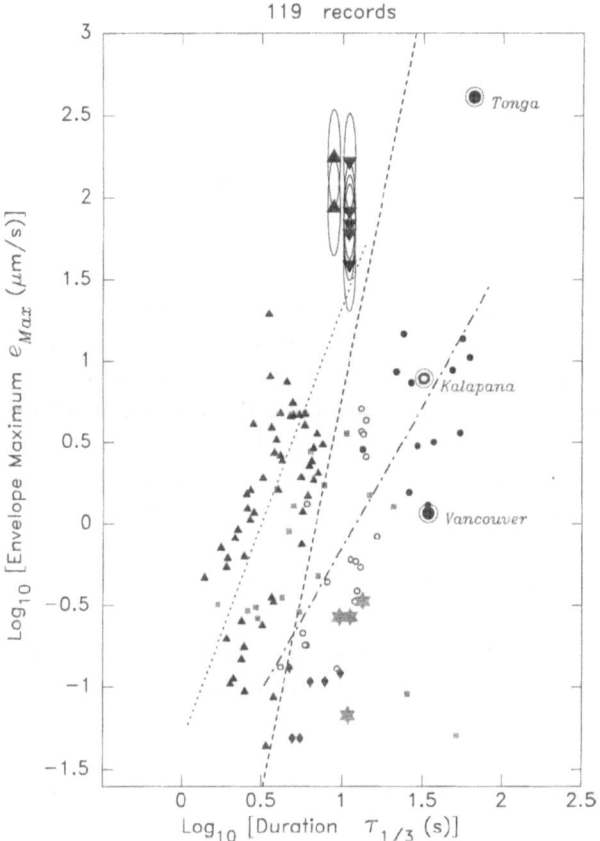

Figure 14

Same as Figure 6, but including the strong sources studied in Section 4. The data points previously used in the digital study are plotted in smaller size. In addition to the symbols used in Figure 6, the downward-pointing triangles are the presumed Vancouver explosions, the stars of David represent the 4 Christmas atmospheric nuclear tests and the circled symbols and labels identify the additional three earthquakes. Error ellipses are shown for the estimated parameters of the confirmed and presumed large-scale explosions. As on Figure 6, the dashed line is the separator (5). In addition, the dotted line represents the regression of the explosion data set (8.E), and the dash-dot line, that of the full earthquake data set (8.S-H); note that their slopes are different.

After regrouping the ten additional data points into the original data set, we computed the following linear regressions for the various types of records:

$$\log_{10} e_{\text{Max}} = 2.67 \log_{10} \tau_{1/3} - 1.34 \qquad (8.\text{E})$$

for the 60 explosion records (E and P);

$$\log_{10} e_{\text{Max}} = 1.45 \log_{10} \tau_{1/3} - 1.44 \qquad (8.\text{S})$$

for the 14 subduction earthquake records (S);

$$\log_{10} e_{\text{Max}} = 2.01 \log_{10} \tau_{1/3} - 2.16 \tag{8.H}$$

for the 19 hotspot earthquake records (H); and

$$\log_{10} e_{\text{Max}} = 1.75 \log_{10} \tau_{1/3} - 1.90 \tag{8.S-H}$$

when combining all 33 earthquake records (S and H).

Finally, we turn to the case of the four atmospheric nuclear tests at Christmas Island in June–July 1962. As shown on Figure 15, the paper records at PPT are unclipped and of sufficient quality to allow hand digitizing after optical magnification of the records. Classical data processing including removal of pen curvature yielded digital time series which were then processed through our standard algorithm. The spectrogram for the test of 10 July 1962 is shown at the bottom of Figure 15. As shown on Figure 14, where the four nuclear tests are plotted as stars of David, T waves from these atmospheric events differ substantially from those of underwater explosions, with all four data points falling on the "earthquake" side of the separator. Their characteristics are most similar to those of small intraplate earthquakes, and also to the later part of missile firing records, but they lack the high-frequency components found in the latter in the early parts of the records, and believed to originate in the underwater part of the firing sequence.

5. Conclusion and Perspective on the $[e_{\text{Max}} : \tau_{1/3}]$ Discriminant

We have developed a discriminant based on the comparison of the maximum amplitude of ground velocity in a T-phase seismogram with the duration of the phase measured at $1/3$ of its maximum. Measurements should be taken on the envelope of the seismogram along the procedure outlined above, and only stations on atolls, or at sites featuring a simple receiver-side conversion process, should be used. The discriminant, shown as the dashed line on Figure 14, successfully separates earthquakes from explosions. Other algorithms, notably those in the frequency domain, have a significant rate of failure, notably for intraplate earthquakes emanating from oceanic hotspot islands. The performance of the discriminant was tested successfully over 3.5 orders of magnitude of ground velocity by analyzing analog records of high-energy underwater explosions from the 1960s.

The proposed discriminant, e_{Max} vs. $\tau_{1/3}$ compares the maximum amplitude of the signal envelope, which is an instantaneous characteristic of its shape, with the duration at $1/3$ of its maximum, which on the contrary, is related to the evolution with time of the signal, and hence, to a form of integral over time. Thus, the basic nature of the discriminant is rooted in the comparison of a higher-frequency parameter with a lower-frequency one, and its apparent success at separating

(N) ATMOS. NUCLEAR TEST, 10 JUL 1962

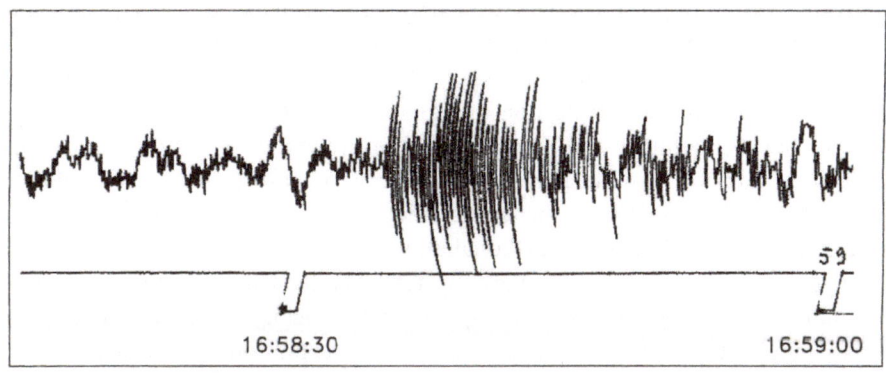

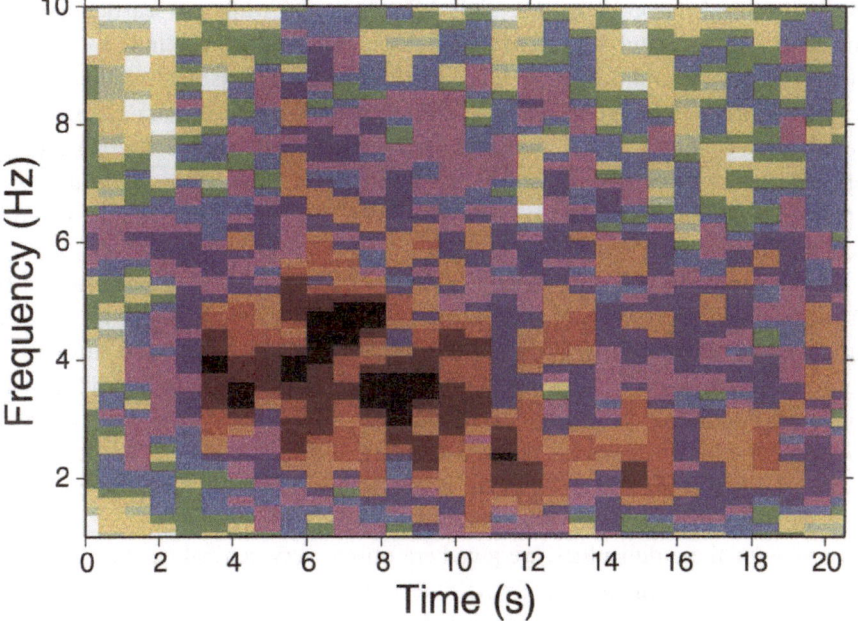

Figure 15

T wave recorded at Papeete from the atmospheric nuclear test at Christmas Island on 10 July 1962. *Top*: Original seismogram. The two tick marks are separated by 30 seconds. *Bottom*: Spectrogram obtained after hand-digitization of the record. Coding as on Figure 2.

earthquakes from explosions should not be surprising. (That it performs better than specific frequency-domain criteria remains intriguing.) Indeed, it shares a certain philosophy with such counterparts in land-based seismology as the time-honored m_b : M_s criterion (MARSHALL and BASHAM, 1972), or even the E/M_0 ratio between energy and moment recently introduced by NEWMAN and OKAL (1998) to identify earthquakes with anomalously slow sources.

It would be desirable to cast the satisfactory performance of $[e_{\text{Max}} : \tau_{1/3}]$, observed over as much as 3.5 orders of magnitude of e, into the framework of a theoretical model of the generation of the T phases by both kinds of sources, and to provide some level of theoretical justification of Equations (8). However, this is a difficult endeavor since it requires the modeling of the growth with source size of four parameters, namely amplitude and time history of earthquakes and explosion sources. There exist models of either source and of the generation of the waves in the relevant medium. In the case of earthquakes, we face the formidable problem of the seismic-to-acoustic conversion, strongly affected by bathymetric features on a scale (a few hundred m) inaccessible to both mapping and modeling. Also, the two kinds of sources may not be directly comparable since an explosion will behave at least initially as a point source in the water, with an initially spherical growth of the wavefront (until trapped by the SOFAR channel), while an earthquake will illuminate a finite segment of conversion slope and thus behave initially as a line source for oceanic sound waves. Under these conditions, it is unclear that the two parameters generally computable from models of either source, namely the seismic ground motion at the source-side conversion shore (earthquake source) and the amplitude of the pressure pulse immediately outside the range of nonlinearity (explosion), can be simply related to the amplitudes of T phases eventually recorded at an on-shore seismic station.

Furthermore, the duration of the T phase is itself an *a priori* combination of the intrinsic duration of the source, the dispersion during propagation in the SOFAR channel, and the complexity of the conversion on the receiver side. The latter can be largely eliminated by considering only atoll stations. Dispersion will become prominent for signals with short source times and thus will more strongly affect explosion signals; furthermore, it could be feasible to effect a dispersion correction before taking the measurement of $\tau_{1/3}$; this improvement to the discriminant is presently being investigated.

In view of these difficulties, we give here only a very general perspective on the nature of the discriminant $[e_{\text{Max}} : \tau_{1/3}]$.

In the case of earthquakes, scaling models relating seismic displacements to earthquake size (e.g., GELLER, 1976) predict interference effects, followed by full saturation of the ground motion at any frequency f when the duration of the source becomes greater than $1/f$. Full saturation for a 5-Hz wave is expected at a seismic moment $M_0 = 10^{22}$ dyn-cm or $m_b = 4.4$ (rather than at the familiar 10^{28} dyn-cm for the Richter magnitude measured at a period 100 times longer). Destructive interference will start about one unit of magnitude below that, meaning that most earthquakes considered in this study will be affected. Hence, the amplitude of seismic motion and consequently of the T phase will grow slower than M_0. Similarly, while the duration of the source would be expected to grow as $M_0^{1/3}$, that of the T phase will be primarily controlled by the duration of excitation of the scatterer at the solid-liquid interface. Consequently, the relationship between e_{Max} and $\tau_{1/3}$ is expected to mimic that existing between the amplitude of strong motion, a, and its duration, D, at a near-field receiver. Such (a, D) relations constitute the basis for the computation of so-called duration magnitudes (LEE et al., 1972). While many variations of such scales have been proposed to adapt them to individual network studies, they generally use an algorithm of the form

$$M_D = b_0 + b_1 \log_{10} D \tag{9}$$

with b_1 varying from 1.5 in Southern California (REAL and TENG, 1973) to 2.8 in Puget Sound (CROSSON, 1972); for example, the preferred value $b_1 = 2.0$ (LEE and STEWART, 1981) would suggest a relationship of the form $\log_{10} a = 2 \log_{10} D$.

The slopes regressed from Equations (8), 2.01 for the mostly small H events, and 1.45 for the generally larger S ones, fall within the general range of used b_1 values, which confirms that the characteristics (amplitude and duration) of the T phases recorded from earthquakes are indeed rooted in those of the seismic ground motion at the seismic-to-acoustic conversion point.

In the case of explosions, we regressed the envelope amplitude e_{Max} as a function of the yield Y (in kg of TNT), as published for 44 sources, and obtained:

$$\log_{10} e_{Max} = 0.77 \log_{10} Y - 2.03 \tag{10}$$

the correlation coefficient being good (87%). The slope of the regression, 0.77, indicates that the amplitude of the recorded T phase grows significantly faster with yield than the amplitude p of the pressure pulse at the source: the latter has been modeled semi-empirically by COLE (1948) and CHAPMAN (1985) as growing like $(Y^{1/3})^{1.13} = Y^{0.38}$. This can be understood through the following argument: The pulse at the source is very narrow, decaying exponentially with a decay time θ_s which grows approximately like $Y^{1/3}(Y^{1/3})^{-0.22} = Y^{0.26}$ (CHAPMAN, 1985). By the time it is recorded with a dominant angular frequency ω at an island station, a T wave will have undergone what amounts to band-pass filtering, and the resulting recorded amplitude should be more directly related to the spectral amplitude $P(\omega)$ of the source pulse than to its time domain maximum, p. For an exponentially decaying signal, and in the limit $\omega\theta_s \ll 1$, we have $P(\omega) = p\theta_s$, which will then grow like $Y^{0.64}$. Of course, this model is very crude, since it assumes a constant ω, and ignores many factors such as the contribution of the secondary pulses (the so-called "bubble" effects), but it gives somes rationale to the observed fast growth of e_{Max} with Y.

As for the duration of the T phase, it is found to correlate poorly with yield (the coefficient being only 68%). This is probably due to the effects of dispersion on the shorter signals generated by explosions. Otherwise, one would expect the duration of the T wave to be controlled by the two characteristic times of the source, the decay time θ_s, and the period of the bubble, θ_B, the latter growing like $Y^{1/3}$. The ratio $0.77/(1/3) = 2.31$ is not far removed from the value (2.67) regressed from the $[e_{Max} : \tau_{1/3}]$ data set (Equation (8.E)).

It is evident that the arguments presented in this section fall short of a full satisfactory theoretical model of the performance of the proposed discriminant. Yet, they provide some level of justification for the observed slopes of regression on Figure 14. When event size is increased, the population of explosions, characterized by the steeper slope (2.67) is bound to separate from the earthquake group featuring an average slope of only 1.75. That this separation is effective in the range of event sizes considered here ($M > 3$ for earthquakes; $Y > 80$ kg for explosions) allows the discriminant to perform efficiently. Assuming that the same scaling laws apply, it is probable that the [$e_{Max} : \tau_{1/3}$] discriminant would fail for smaller events, in the yield range 1–10 kg and in the magnitude range 1–2.

Acknowledgments

This research is supported by the Defense Threat Reduction Agency of the Department of Defense, under Grant DSWA01-98-1-0007, and by Commissariat à l'Energie Atomique (France). The paper was improved through the comments of Jeffrey Stevens and another reviewer.

REFERENCES

ANDO, M. (1979), *The Hawaii Earthquake of November 29, 1975: Low-dip Angle Faulting due to Forceful Injection of Magma*, J. Geophys. Res. *89*, 7616–7626.

ANONYMOUS (1989), *Announced United States Nuclear Tests, July 1945 through December 1988*, Office of External Affairs, U.S. Dept. of Energy, 64 pp.

CHAPMAN, N. R. (1985), *Measurement of Waveform Parameters of Shallow Explosive Charges*, J. Acoust. Soc. Am. *78*, 672–681.

COLE, R. H., *Underwater Explosions* (Princeton Univ. Press, Princeton, N.J., 1948), 437 pp.

CROSSON, R. S. (1992), *Small Earthquakes, Structure, and Tectonics of the Puget Sound Region*, Bull. Seismol. Soc. Am. *62*, 133–1171.

DUENNEBIER, F. K., and JOHNSON R. H. (1967), *T-phase Sources and Earthquake Epicenters in the Pacific Basin*, Hawaii Inst. Geophys. Rept. 67–24, 100 pp., Honolulu.

DZIEWONSKI, A. M., EKSTRÖM, G., and SALGANIK, M. P. (1996), *Centroid-moment Tensor Solutions for January–March 1995*, Phys. Earth Planet. Inter. *95*, 147–157.

DZIEWONSKI, A. M., EKSTRÖM, G., and SALGANIK, M. P. (1997a), *Centroid-moment Tensor Solutions for October–December 1995*, Phys. Earth Planet. Inter. *101*, 1–12.

DZIEWONSKI, A. M., EKSTRÖM, G., and SALGANIK, M. P. (1997b), *Centroid-moment Tensor solutions for January–March 1996*, Phys. Earth Planet. Inter. *102*, 1–9.

DZIEWONSKI, A. M., EKSTRÖM, G., and SALGANIK, M. P. (1997c), *Centroid-moment Tensor Solutions for April–June 1996*, Phys. Earth Planet. Inter. *102*, 11–20.

DZIEWONSKI, A. M., EKSTRÖM, G., MATERNOVSKAYA, N. N., and SALGANIK, M. P. (1997d), *Centroid-moment Tensor Solutions for July–September 1996*, Phys. Earth Planet. Inter. *102*, 133–143.

DZIEWONSKI, A. M., EKSTRÖM, G., and MATERNOVSKAYA, N. (1998), *Centroid-moment Tensor Solutions for October–December 1996*, Phys. Earth Planet. Inter. *105*, 95–108.

DZIEWONSKI, A. M., EKSTRÖM, G., and MATERNOVSKAYA, N. (1999a), *Centroid-moment Tensor Solutions for April–June 1997*, Phys. Earth Planet. Inter. *112*, 1–9.

DZIEWONSKI, A. M., EKSTRÖM, G., and MATERNOVSKAYA, N. (1999b), *Centroid-moment Tensor Solutions for July–September 1997*, Phys. Earth Planet. Inter., in press.

GELLER, R. J. (1976), *Scaling Relations for Earthquake Source Parameters and Magnitudes*, Bull. Seismol. Soc. Amer. *66*, 1501–1523.

LEE, W. H. K., and STEWART, S. W., *Principles and Applications of Microearthquake Networks*, Adv. Geophys. Suppl. 2, 293 pp. (Academic Press, New York, 1981).

LEE, W. H. K., BENNETT, R. E., and MEAGHER, K. L. (1972), *A Method of Estimating Magnitude of Local Earthquakes from Signal Duration*, U.S. Geol. Surv. Open File Rept., 28 pp.

LINEHAN, J. (1946), *Earthquakes in the West Indian Region*, Trans. Am. Geophys. Un. *21*, 229–232.

LUNDGREN, P. R., and OKAL, E. A. (1988), *Slab Decoupling in the Tonga Arc: The June 22, 1977 Earthquake*, J. Geophys. Res. *93*, 13,355–13,366.

MA, K.-F., KANAMORI, H., and SATAKE, K. (1999), *Mechanism of the 1975 Kalapana, Hawaii Earthquake Inferred from Tsunami Data*, J. Geophys. Res. *104*, 13,153–13,167.

MARSHALL, P. D., and BASHAM, P. W. (1972), *Discrimination between Earthquakes and Underground Explosions Using an Improved M_s Scale*, Geophys. J. Roy. astr. Soc. *28*, 431–458.

NAVA, F. A., NÚÑEZ-CORNÚ, F., CÓRDOBA, D., MENA, M., ANSORGE, J., RODRÍGUEZ, M., BANDA, E., MÜLLER, S., UDÍAS, A., GARCI'A-GARCÁ, M., CALDERÓN, G., and the MEXICAN WORKING GROUP FOR DEEP SEISMIC PROFILING, (1988), *Structure of the Middle America Trench in Oaxaca, Mexico*, Tectonophy. *154*, 241–251.

NEWMAN, A. V., and OKAL, E. A. (1998), *Teleseismic Estimates of Radiated Seismic Energy: The E/M_0 Discriminant for Tsunami Earthquakes*, J. Geophys. Res. *103*, 26,885–26,898.

OKAL, E. A. (1989), *A Theoretical Discussion of Time-domain Magnitudes: The Prague Formula for M_s and the Mantle Magnitude M_m*, J. Geophys. Res. *94*, 4194–4204.

OKAL, E. A., and TALANDIER, J. (1986), *T-wave Duration, Magnitudes and Seismic Moment of an Earthquake; Application to Tsunami Warning*, J. Phys. Earth *34*, 19–42.

PISERCHIA, P.-F., VIRIEUX, J., RODRIGUES, D., GAFFET, S., and TALANDIER, J. (1998), *Hybrid Numerical Modeling of T-wave Propagation: Application to the Midplate Experiment*, Geophys. J. Intl. *133*, 789–800.

RAVET, J. (1940), *Remarques sur quelques enregistrements d'ondes à très courte période au cours de tremblements de terre lointains à l'Observatoire du Faiere, Papeete, Tahiti*, Sixth Pacific Sci. Congress, vol. 1, pp. 127–130.

REAL, C. R., and TENG, T.-L. (1973), *Local Richter Magnitude and Total Signal Duration in Southern California*, Bull. Seismol. Soc. Am. *63*, 1809–1827.

TALANDIER, J. (1993), *French Polynesia Tsunami Warning Center (CPPT)*, Natural Hazards *7*, 237–256.

TALANDIER, J., and OKAL, E. A. (1982), *Crises sismiques au volcan Macdonald (Océan Pacifique Sud)*, C.R. Acad. Sci. Paris, Sér. II. *295*, 195–200.

TALANDIER, J., and OKAL, E. A. (1984a), *The Volcanoseismic Swarms of 1981–1983 in the Tahiti-Mehetia Area, French Polynesia*, J. Geophys. Res. *89*, 11,1216–11,234.

TALANDIER, J., and OKAL, E. A. (1984b), *New Surveys of Macdonald Seamount, Southcentral Pacific, Following Volcanoseismic Activity, 1977–1983*, Geophys. Res. Lett. *11*, 813–816.

TALANDIER, J., and OKAL, E. A. (1987a), *Seismic Detection of Underwater Volcanism: The Example of French Polynesia*, Pure appl. geophys. *125*, 919–950.

TALANDIER, J., and OKAL, E. A. (1987b), *Crustal Structure in the Tuamotu and Society Islands, French Polynesia*, Geophys. J. Roy. astr. Soc. *88*, 499–528.

TALANDIER, J., and OKAL, E. A. (1998), *On the Mechanism of Conversion of Seismic Waves to and from T Waves in the Vicinity of Island Shores*, Bull. Seismol. Soc. Am. *88*, 621–632.

WEIGEL, W., *Bericht über die SONNE-Expedition SO65-2, Papeete-Papeete, 7.-28. Dez. 1989* (Universität Hamburg, Institut für Geophysik, 1990).

(Received June 30, 1999, revised November 8, 1999, accepted November 15, 1999)

 To access this journal online:
http://www.birkhauser.ch

Pure appl. geophys. 158 (2001) 605–626
0033–4553/01/030605–22 $ 1.50 + 0.20/0

▌Pure and Applied Geophysics

Analysis of Russian Hydroacoustic Data for CTBT Monitoring

MARIANA ENEVA,[1] JEFFRY L. STEVENS,[1] BORIS D. KHRISTOFOROV,[2]
JACK MURPHY,[3] and VITALY V. ADUSHKIN[2]

Abstract — As part of a collaborative research program for the purpose of monitoring the Comprehensive Nuclear-Test-Ban Treaty (CTBT), we are in the process of examining and analyzing hydroacoustic data from underwater explosions conducted in the former Soviet Union. We are using these data as constraints on modeling the hydroacoustic source as a function of depth below the water surface. This is of interest to the CTBT because although even small explosions at depth generate signals easily observable at large distances, the hydroacoustic source amplitude decreases as the source approaches the surface. Consequently, explosions in the ocean will be more difficult to identify if they are on or near the ocean surface. We are particularly interested in records featuring various combinations of depths of explosion, and distances and depths of recording.

Unique historical Russian data sets have now become available from test explosions of 100-kg TNT cast spherical charges in a shallow reservoir (87 m length, 25 m to 55 m width, and 3 m depth) with a low-velocity air-saturated layer of sand on the bottom. A number of tests were conducted with varying water level and charge depths. Pressure measurements were taken at varying depths and horizontal distances in the water. The available data include measurements of peak pressures from all explosions and digitized pressure-time histories from some of them. A reduction of peak pressure by about 60–70% is observed in these measurements for half-immersed charges as compared with deeper explosions. In addition, several peak-pressure measurements are also available from a 1957 underwater nuclear explosion (yield < 10 kt and depth 30 m) in the Bay of Chernaya (Novaya Zemlya).

The 100-kg TNT data were compared with model predictions. Shockwave modeling is based on spherical wave propagation and finite element calculations, constrained by empirical data from US underwater chemical and nuclear tests. Modeling was performed for digitized pressure-time histories from two fully-immersed explosions and one explosion of a half-immersed charge, as well as for the peak-pressure measurements from all explosions carried out in the reservoir with water level at its maximum (3 m). We found that the model predictions match the Russian data well.

Peak-pressure measurements and pressure-time histories were simulated at 10 km distance from hypothetical 1-kt and 10-kt nuclear explosions conducted at various depths in the ocean. The ocean water was characterized by a realistic sound velocity profile featuring a velocity minimum at 700 m depth. Simulated measurements at that same depth predict at least a tenfold increase in peak pressures from explosions in the SOFAR channel as compared with very shallow explosions (e.g., ~3 m depth).

Work supported by Defense Threat Reduction Agency Contract DSWA01-97-C-0166

[1] Maxwell Technologies, Systems Division, 9210 Park Court, San Diego, California 92123, USA. E-mail: meneva@maxwell.com; stevens@maxwell.com
[2] Institute for Dynamics of the Geospheres, Russian Academy of Sciences, Leninsky Prospect, 38 Korpus 6, Moscow 117334, Russia.
[3] Maxwell Technologies, Systems Division, 11800 Sunrise Valley Drive, Suite 1212, Reston, VA 20191-5309, USA.

The observations and the modeling results were also compared with predictions calculated at the Lawrence Livermore National Laboratory using a different modeling approach. All results suggest that although the coupling is reduced for very shallow explosions, a shallow 1-kt explosion should be detectable by the IMS hydroacoustic network.

Key words: Hydroacoustic, Russian tests, modeling, shallow underwater explosions.

1. Introduction

As part of a collaborative research program for the purpose of monitoring the Comprehensive Nuclear-Test-Ban Treaty (CTBT), we are in the process of examining and analyzing hydroacoustic data from underwater explosions conducted in the former Soviet Union. The CTBT hydroacoustic network consists of 11 stations that monitor the oceans for underwater explosions, as well as for atmospheric explosions conducted close to the ocean surface. The reason such a small network can monitor the whole world is that hydroacoustic waves propagate very efficiently in the acoustic waveguide known as the SOFAR channel (e.g., WALKER *et al.*, 1992). Since a significant portion of the energy from underwater explosions couples to this channel as acoustic waves, even kilogram-sized explosions at depth generate signals easily observable at large distances. For example, STEVENS *et al.* (2001) discuss hydroacoustic signals from four-pound charges detonated off the coast of San Francisco that were recorded as far away as Wake Island, a distance of about 7,000 km.

The coupled energy is the energy transferred from the source to the water, manifested as a point-source explosion generating an initially spherically divergent underwater shockwave. The amount of energy coupled to the water depends strongly on the explosion depth. Of interest to the CTBT are shallow and surface sources, as the explosion energy from such tests would be significantly decoupled from the ocean and could be much more difficult to detect. In light of the above, we are particularly interested in records featuring various combinations of depths of explosion, and distances and depths of recording. Historical Russian hydroacoustic data have become available recently that consist of a number of such measurements. The usefulness of these data lies in the possibility of understanding better the coupling for shallow and surface explosions, and how they differ from deeper, fully coupled explosions. Thus we are using these data as constraints on modeling the hydroacoustic source as a function of depth below the water surface. The focus in this work is on analyzing and modeling uniquely comprehensive data from smaller-scale explosions in a shallow reservoir, but several measurements from an underwater nuclear explosion are also examined. Scaling rules can be applied to relate smaller explosions to the nuclear explosions of interest to the CTBT. Additional hydroacoustic data from Russian nuclear underwater explosions are becoming available at time of writing, to be analyzed in a future work.

Understanding the effects of underwater explosions is largely based on measurements and modeling of the related shockwaves and bubble pulsations (e.g., NRC, 1997). During the early stages of an underwater explosion, the materials of the device attain high temperatures and pressures. Energy acquired by these materials heats and compresses the surrounding water. This mechanism forms a hydrodynamic compression wave that moves outward at a faster rate than the material it engulfs. An almost instantaneous increase in pressure occurs at the shock front, while the pressure decreases more gradually behind it. This is the primary shockwave. As the shock front moves away from the source region, energy dissipated as heat raises the temperature of the ambient water, with the largest temperature increase occurring near the center of the explosion. This causes both vaporization and dissociation of water in the explosion center, while at greater distances the water is vaporized into steam, forming an expanding steam bubble. If the explosion is deep enough, several cycles of bubble expansions and contractions may occur, resulting in progressively weaker pulses. Both the small-scale Russian experiments and the underwater nuclear explosions were conducted in shallow water, so that the hydroacoustic data analyzed here do not include bubbles and feature only shockwave parameters (peak pressure, duration and impulse) and pressure-time histories. The data clearly demonstrates that the underwater shockwaves from explosions at or near the water surface depend strongly on the amount of energy coupled to the water in the source region and peak pressures are significantly diminished by comparison with deeper explosions (by 60–70%).

The modeling in the present work uses a recent version of a numerical code known originally as REFM (Reflection and Refraction in Multi-Layered Ocean/ Ocean Bottoms) and later as REFMS. This code has been evolving for more than twenty years now (BRITT et al., 1991; BRITT, 1985). It is based on the Cagniard solution for expanding spherical waves in a layered medium (CAGNIARD et al., 1962), with the characterization of boundary interactions improved by finite difference modeling. REFMS modeling includes surface and bottom reflections and refraction due to sound velocity gradients. Although shallow-water and near-surface explosions are known to be particularly difficult to model accurately, many of the measurements in the Russian experiments are predicted very well using this modeling. REFMS was also used to simulate pressure-time histories and maximum peak pressures from 1-kt and 10-kt nuclear sources detonated in the ocean, in the presence of a realistic sound-velocity profile.

A different type of modeling of the near-surface hydroacoustic source has been employed by a group at the Lawrence Livermore National Laboratory (LLNL), linking two computational codes (CLARKE et al., 1995). The first code, CALE, is a LLNL hydrodynamic code used for the strong-shock calculations. The second code, NPE, is a Naval Research Laboratory (NRL) code used to model weak-shock propagation. These authors estimated the coupling factor by calculating the energy of the shockwaves at a 10-km range from a 1-kt nuclear source for a variety of explosion depths below the ocean surface and heights above it. In the following analysis, we compare relevant results from the LLNL work (CLARKE *et al.*, 1995) with the Russian observations and the REFMS modeling performed in our study.

2. Data

Unique historical Russian data have become available recently, among them a variety of hydroacoustic measurements. These include data from 100-kg TNT explosions in a shallow reservoir, explosions of 136-kg bombs in the deep waters of the Sea of Okhotsk, detonation of a 500-m long cord at depth 1 m in the Black Sea, and some underwater nuclear explosions in Novaya Zemlya. The focus here is on the comprehensive data set from the 100-kg TNT explosions in shallow water. Preliminary modeling of data from an underwater nuclear explosion is also discussed.

2.1. 100-kg TNT Explosions in Shallow Water

Twenty-nine explosion experiments were conducted in a shallow reservoir in the former Soviet Union (KOZACHENKO and KHRISTOFOROV, 1970; KOROBEINIKOV and KHRISTOFOROV, 1976). The radius of the 100-kg TNT cast spherical charges was about 0.25 m. The dimensions of the reservoir were as follows: 87 m length, 3 m depth, and 55 m to 25 m width from top to bottom. Various experimental setups were used, featuring a charge depth from surface (0 m, or half-immersed) to 2.75 m (charge on the bottom) and water level from 0.25 m to 3 m. The bottom of the reservoir consisted of a 1-m thick layer of air-saturated sand with a very low sound velocity of 270 m/s; that is, the effect of this layer would not be too different from that of the air above the water surface. Water was pumped out after each explosion, the crater was leveled out and water was pumped in again for the next explosion. Figure 1 schematically represents the experimental setup of the explosions.

Measurements from these explosions were made with piezoelectric gauges suspended at various distances (7.5 m, 15 m, 22.5 m, and 30 m) and depths (0.25 m to 2.75 m) in the water. Shock-parameter (peak pressure, pulse duration, and specific impulse) measurements are now available from all 100-kg TNT explosions. The records of pressure-time and impulse-time histories have been already digitized and made available to us as well.

In the present work we utilize only measurements from 12 explosions performed in a full reservoir (water level of 3 m) and do not consider data from the remaining 17 explosions in shallower water. In particular, we use peak-pressure measurements at various distances and depths, as well as 26 digitized pressure records from two fully immersed explosions (1-m charge depth) and 14 such records from one half-

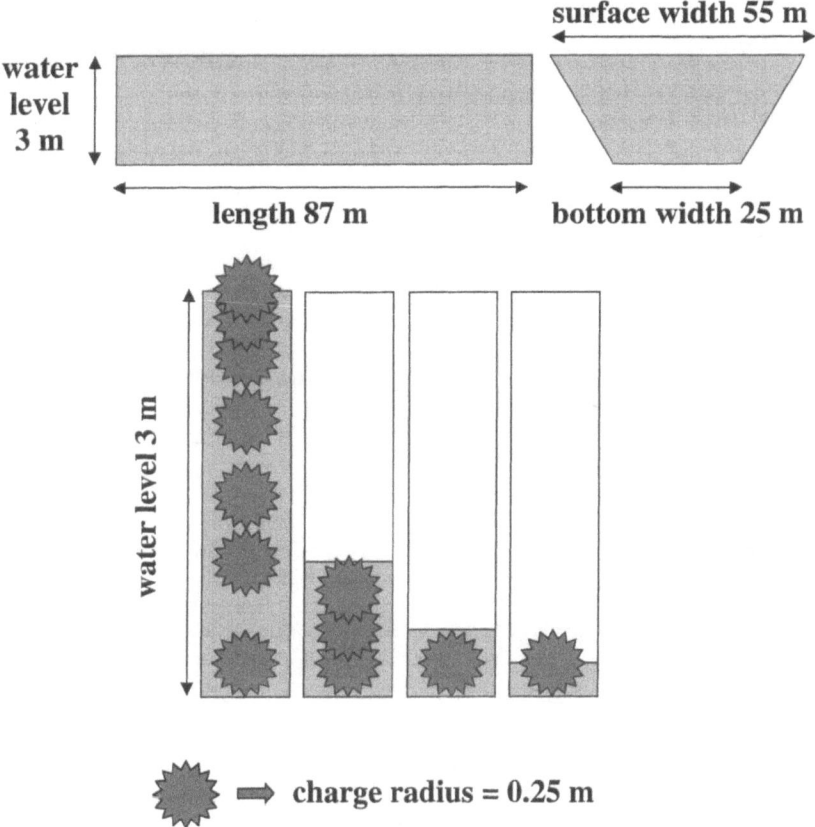

Figure 1

Schematic representation of the reservoir and the experiments with underwater explosions. Top – cross sections of the reservoir (dimensions not to scale). Bottom – 29 explosions were carried out featuring various combinations of water levels (0.25 m to 3 m) and charge depths (0 m to 2.75 m).

immersed charge (0-m charge depth). Figures 2 and 3 below show the available pressure-time histories from fully- and half-immersed charges, respectively, along with modeling results to be discussed in the next section.

Comparing the peak pressures measured at depth of 1.5 m from half-immersed and fully-immersed explosions shows that close to a 60% reduction is observed at a distance of 15 m and 70% at 30 m (measurements at distances 7.5 m and 22.5 m are not available for the half-immersed charges). These percentages were calculated from the relative peak-pressure differences at the two distances, (147.5–60.5)/147.5 and (66–19.3)/66, respectively. The sensor depth of 1.5 m (mid-pool) was chosen in the above comparison because it can be assumed that measurements at this depth are least affected by surface and bottom reflections and because the largest peak pressures were usually observed at these intermediate sensor depths.

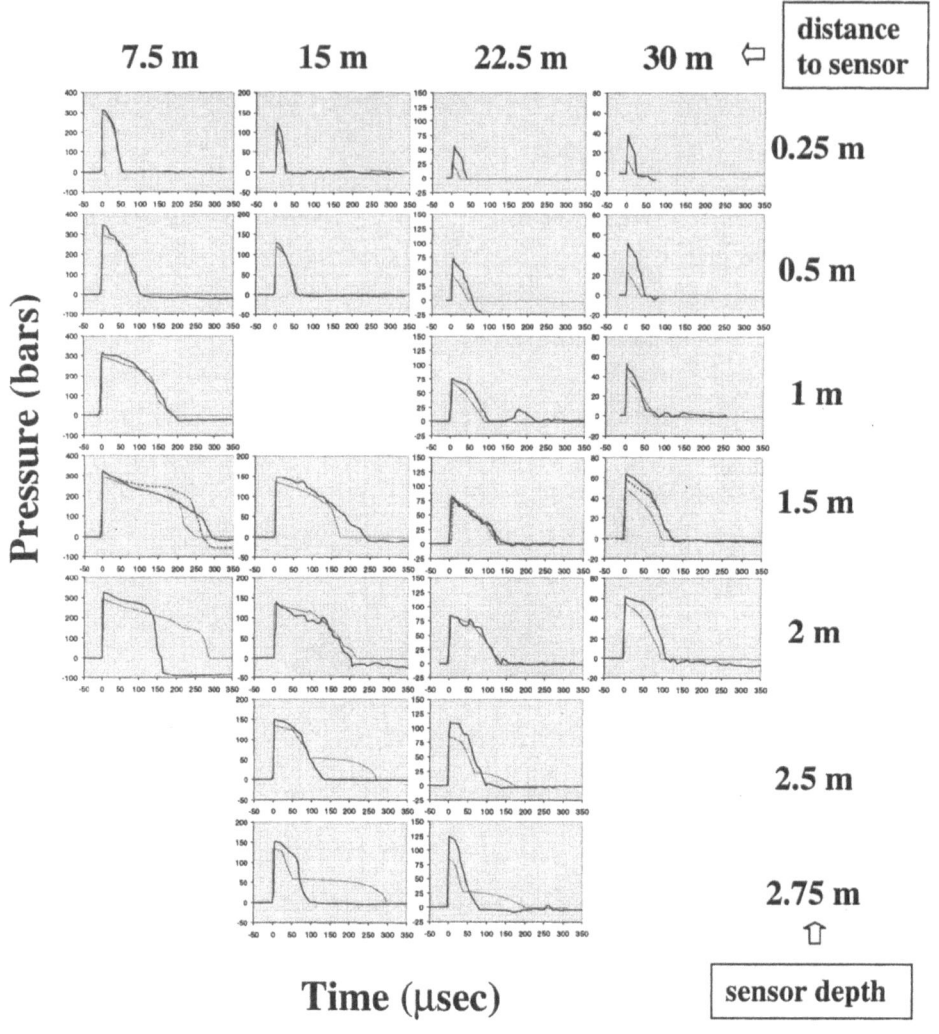

Figure 2
Observed and modeled pressure-time histories for fully-immersed charges (charge depth 1 m, water level
3 m). Modeled – gray solid lines; observed – black solid lines, with dashed lines indicating additional
measurements at sensor depth 1.5 m.

At present, there is virtually no specific information about the errors involved in
these measurements. Errors of 8%, 10%, and 18% were cited in the archives for
the measurements of peak pressure, specific impulse, and energy, respectively, in
one case only. This lack of error estimates is in contrast to the error control
apparently intended in the test design. It is known that several gauges were
suspended in water at identical distances and depths to take measurements from

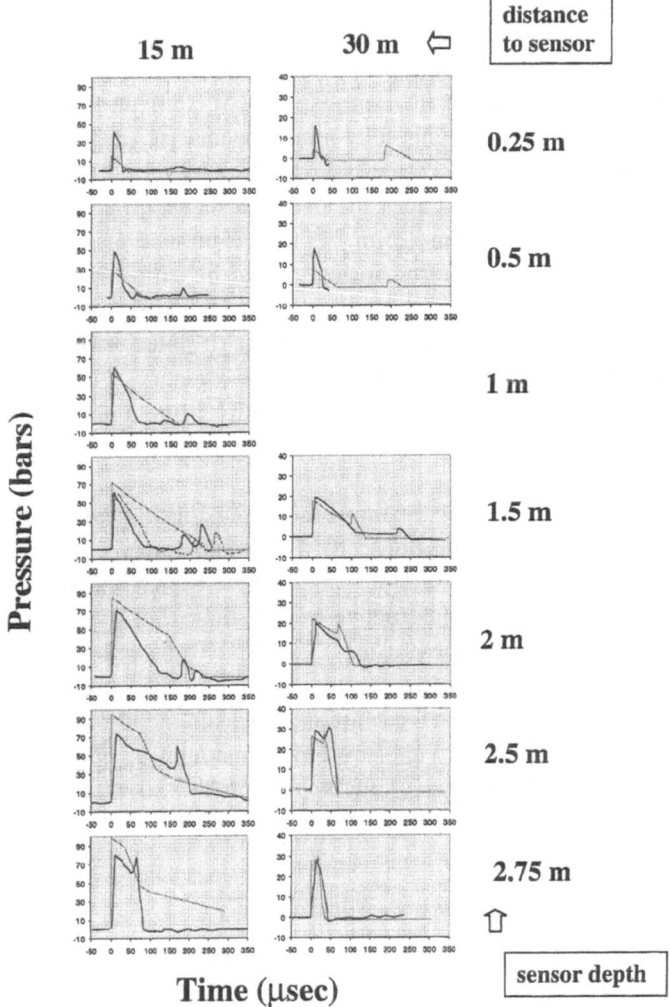

Figure 3
Observed and modeled pressure-time histories for a half-immersed charge (charge depth 0 m, water level 3 m). Modeled – gray solid lines; observed – black solid lines, with dashed lines indicating additional measurements at sensor depth 1.5 m.

identical explosions. In addition, many of the explosion configurations were repeated in several tests. The former would provide information on measurement errors *per se*, while the latter would add differences due to difficulties in repeating the explosion configurations exactly (e.g., discrepancies in measurements of distances and depths in field conditions, changes from test to test in sound speed of the bottom, etc.). Differences in measurements from different explosions are therefore expected to be larger on average than the differences between measurements from the same experiments.

We attempted to make some representative estimates of the differences among peak-pressure measurements for the same experiment configurations, whenever these are available. For the case involving fully-immersed charges (explosion depth 1 m) in a full pool (water level 3 m), it was reassuring to observe that none of the relative differences among two different measurements for each of three combinations of sensor distance and depth (7.5 m–1.5 m, 7.5 m–0.25 m, and 15 m–0.25 m) exceeded 1.3%. Differences in pulse duration turned out to be rather large by comparison, from 6% to 28%. Each of these repeated measurements was made in identical tests. It is not known how representative these three isolated estimates are for the remaining peak-pressure measurements in this category. No observations are available at identical sensor distances and depths from different fully-immersed explosions, but it is reasonable to expect larger discrepancies among measurements in this case.

We also estimated discrepancies among measurements of peak pressure and pulse duration from 9 other configurations, all with water levels 1 m and 0.25 m; i.e., charges were detonated in very shallow water. These explosions are not analyzed in the present work, but it is still instructive to note that peak-pressure measurements from identical explosions differed at most by 10%, while differences among some tests exceeded 20%. Thus compared to the detonation of charges in full reservoir, larger discrepancies were observed among measurements from shallow-water explosions. This is likely due to more prominent nonlinear effects, and therefore, stronger sensitivity to differences among measurement and explosion conditions in very shallow water.

2.2. Underwater Nuclear Explosion in Novaya Zemlya

Three underwater nuclear explosions were carried out in the Bay of Chernaya, Novaya Zemlya (70.70°N, 54.67°E) in the period 1955–1961. These explosions were conducted in shallow water, as the bottom in the Bay of Chernaya is 60 m to 100 m deep. The purpose was to test torpedo launches with various nuclear charges and to study their effects on military equipment, such as ships, submarines, and buildings on the coast. The site of these explosions is characterized by low water exchange between the Bay of Chernaya and the Barents Sea as they are connected only by a narrow strait. Thus there was relatively low radioactive contamination of the Arctic Ocean and the continental shore from these explosions.

The hydroacoustic data available at present consist of several peak-pressure measurements (Fig. 4) at a distance of 235 m from the second of the three underwater tests, conducted on October 10, 1957. The reported yield and depth of explosion were 10 kt and 30 m, respectively (*USSR Nuclear Weapons Tests und Peaceful Nuclear Explosions 1949 through 1990*, 1996). For this test, a torpedo was

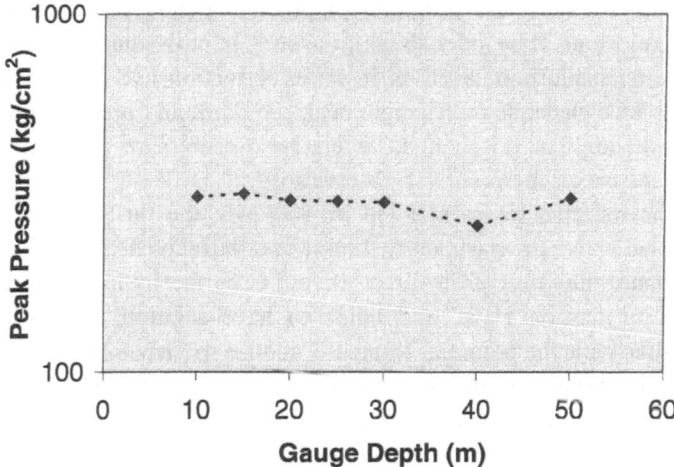

Figure 4
Peak-pressure measurements at a distance of 235 m and various sensor depths from an underwater nuclear explosion in the Bay of Chernaya, Novaya Zemlya (October 10, 1957), reported with an yield of 10 kt and depth 30 m.

launched from a submarine at a periscope depth. Its target was an old ship at a distance of 10 km. Shockwave parameters were measured using mechanic and tourmaline piezoelectric gauges and piston impulse meters hung from boats at various distances and depths. In order to assure more precise measurements, up to six sensors were used at any given location. A measurement accuracy of about 15% is mentioned in the archival materials. The decrease in peak-pressure seen at sensor depth of 40 m is slightly larger than this error estimate, and if real, may be due to some shadowing effect. No information is available for the sound-velocity profile of water and bottom at the time of the explosion, so that it is difficult to distinguish between a real effect and a measurement error. The average of the peak-pressure measurements from this underwater nuclear explosion ($\sim$300 bars) is in the range of measurements from the 100-kg TNT fully-immersed explosion experiments at a distance of 7.5 m (see Fig. 2). We will expand on this similarity later in the paper (Section 4.2 and Fig. 10).

3. Modeling

The modeling of the shockwaves in the present work was performed using one of the codes, Underwater_Shock, included in the so-called DNA (former Defense Nuclear Agency) Computational Aids (e.g., STEPHENS and KELLY, 1995). It represents a modified and simplified Windows version of the ray-tracing code REFMS, that continues to undergo various improvements (BRITT et al., 1991; BRITT,

1985). REFMS is a computer program for predicting shockwave parameters from underwater explosions. It includes all major aspects of near-source wave propagation from underwater explosions, such as in-water refraction and bottom and surface reflection. We used the code to calculate peak pressures and pressure-time histories. The term "pressure" in this context refers to overpressure as compared to the hydrostatic pressure at the points of observation.

Among the reflected waves handled by REFMS, the surface-reflected and the bottom-reflected waves are particularly important, with an effect comparable to and sometimes greater than that of the direct wave. The surface-reflected wave represents a rarefaction or tension wave, and tends to have a cutoff effect on the direct shockwave, thus with the potential to greatly reduce it. Depending on the two path lengths, the pulse decay rate, and the gauge depth, the cutoff may be partial, complete, or cavitation-limited. The bottom reflections can have both positive and negative components depending on the properties of the bottom layers. Surface-bottom and surface-bottom-surface-reflected waves are also calculated in the code. Higher-order multiple-reflection paths are relatively insignificant because of their greater attenuation and longer path lengths. If gradients are present in the water sound velocity, the code also handles refraction, using the similarity between refracted shockwaves and refracted sound waves. Refraction is calculated through the consideration of up to 150 discrete layers in the water and the bottom, each having a constant sound velocity and density. Regions of substantial focusing (caustics) or shadowing of pressure are predicted with significant accuracy, featuring up to tenfold changes in peak pressure for HE (high-explosive) yields, and up to fivefold changes for nuclear yields, as compared with unbounded homogeneous water.

The computational tools included in the REFMS modeling include two main features: (1) use of spherical wavefronts and (2) finite-element calculations taking into account deviations from acoustic propagation. The former is based on the acoustic spherical wave reflection theory formulated by CAGNIARD *et al.* (1962) and assures more realistic predictions of the pressure histories than if only plane waves are used. The second feature is needed because finite-amplitude effects, not considered in the regular acoustic approximation, become more important with increasing incident pressures and incidence angles (e.g., greater than 170 bars and 85°, respectively). In an acoustic approximation, it is assumed that both the direct and surface-reflected shockwaves propagate at the speed of sound waves with infinitesimal amplitudes in undisturbed ambient water, and that the water particle velocity is negligibly small relative to the propagation velocity. While this approximation is valid at relatively short times after the underwater explosion (hence, small horizontal ranges), the reflected wave travels in fact faster than the direct wave in the region close to the interface (water-air in this case), and both waves travel faster than the sound speed in water. Further from the explosion this discrepancy becomes quite apparent. Thus the reflected wavefront tends to close in on the direct wavefront, eventually catching up to and merging with it. The region of merged wavefronts extends from the surface down to a depth that increases as the waves propagate further. It is known as the region of nonregular reflection, in which neither the peak pressures, nor the pressure-time histories, are predictable by the acoustic approximation method, but only by finite-amplitude computations. The peak pressures in the region of nonregular reflection are lower than that of a direct shockwave in unbounded water and presumably decrease to zero at the surface just above this region. For gauge locations below this region, the peak pressures may be predictable by the acoustic approximation method, but finite-amplitude computations are still needed to predict the entire pressure-time history.

The finite-amplitude surface reflection calculations in REFMS have been validated with a series of HE experiments. Predictions were generally very good, with the exception of some discrepancies at the shallowest gauges. In terms of refraction, the code is in good agreement with data from underwater nuclear explosion tests (Wigwam, Wahoo, Umbrella, and Swordfish), as well as from HE experimental series performed on a laboratory scale, in flooded quarries, and in the ocean (e.g., BRITT et al., 1991; BROCKHURST et al., 1961).

Near-surface explosions are of particular importance for CTBT purposes, yet they are most difficult to model accurately because of strong nonlinear or finite-amplitude effects. In terms of what explosion depths can be considered as shallow-submerged, the usual assumption is a depth smaller than scaled depth $1 \text{ m/kt}^{1/3}$. This translates into less than 5 cm for the Russian experiments, which do indeed include half-immersed charges (0 m depth). However, because of the charge radius of 25 cm, the tests featuring charge depths of 25 cm also fall in this category. On the other hand, it is expected that full coupling of energy to water is approached at scaled depths of about $5 \text{ m/kt}^{1/3}$, which translates into about 24 cm for the reservoir experiments. Again, taking into account the charge radius, explosion depths of 0.5 m are a border case, but charges with depths of 1 m and deeper fall in this category, thus justifying the term "fully-immersed" we use throughout the paper for these explosions.

4. Modeling Results

4.1. 100-kg TNT Explosions in Shallow Water

The observed pressure-time histories from the explosions of fully-immersed charges in Figure 2 are shown together with the predicted signals using the REFMS modeling. The modeling was done assuming a constant sound velocity in water of 1500 m/s (i.e., no velocity gradients and hence, no in-water refraction). All other parameters in the modeling matched exactly the configuration of the explosions (depth 1 m, water level 3 m) and the sensors (depth 0.25 m to 2.75 m, horizontal distance from explosion 7.5 m to 30 m), with one exception. Because of limitations of the used version of the code, the bottom layer of 1-m thickness could not be modeled with the original very low sound velocity (270 m/s). This low velocity was likely used so as to not destroy the reservoir in these experiments, and its effect on the measurements was comparable to that of the air above the water surface. The minimum allowed velocity in the modeling (1220 m/s), is higher than the original one. Although it matches qualitatively the real low-velocity bottom (i.e., sound velocity in the bottom is lower than the sound velocity in the water), this discrepancy is the likely cause of the differences between predicted and observed signals at larger sensor depths in Figure 2 (e.g., see the pressure-time histories for sensor depth 2 m at distance 7.5 m and for sensor depths 2.5 m and 2.75 m). As to the large discrepancies seen at smaller sensor depths (0.25 m and 0.5 m) at larger distances (22.5 m and 30 m), they may be due to an "overinterpretation" of the nonregular surface reflection, with the observations not supporting the predicted spatial extent of such a region in this case. Uncontrollable experiment variations (e.g., water-surface waves, refraction due to thermal gradients in the upper layers of water, etc.) may also cause such discrepancies. Predictions at larger distances are generally worse than the ones at smaller distances. Large-distance predictions show the smallest differences from the observations at intermediate sensor depths (1 m and 1.5 m), likely because the effects of the surface and the bottom are least significant at these depths. The predicted signals for the remaining combinations of distances and sensor depths in Figure 2 match the observed ones very well, even in what can be considered as essentially the source region of the explosions (small distances of 7.5 m and sensor depths smaller than 2 m).

Figure 3 shows observed and predicted pressure-time histories from a half-immersed explosion. Due to the larger role of nonlinear effects, and in particular, the not-fully-understood change in coupling with charge depth, such explosions are substantially more difficult to model and larger discrepancies are to be expected. Model parameters followed the test configuration in the same manner as for the fully-immersed explosions above, with one additional exception. Charge depth could not be put at 0 m; a depth of 3 cm was used instead, as the minimum allowed by this version of the code. It is assumed that no significant problems are caused by this discrepancy. Given the difficulty of modeling surface explosions, Figure 3 indicates a good agreement between observed and predicted pressure-time histories, with similar problems to the ones revealed by Figure 2. These include possible overinterpretation of the region of nonregular surface reflection at small depths (0.25 m and 0.5 m) and problems at larger sensor depths due to the mismatch between the modeled and the real sound velocities in the bottom.

Figure 5 depicts the observed and the predicted relationship between peak pressures and sensor depth for the case of a half-immersed explosion (0-m depth). It is apparent that the modeling is generally the best for larger sensor depths at the largest horizontal distance (30 m). The mismatch between the real and model sound velocities in the bottom apparently cause some discrepancy at larger sensor depths closer to the source (15 m), but not further (30 m). For sensor depths close to the surface the modeling probably exaggerates the effect of nonregular reflection and the

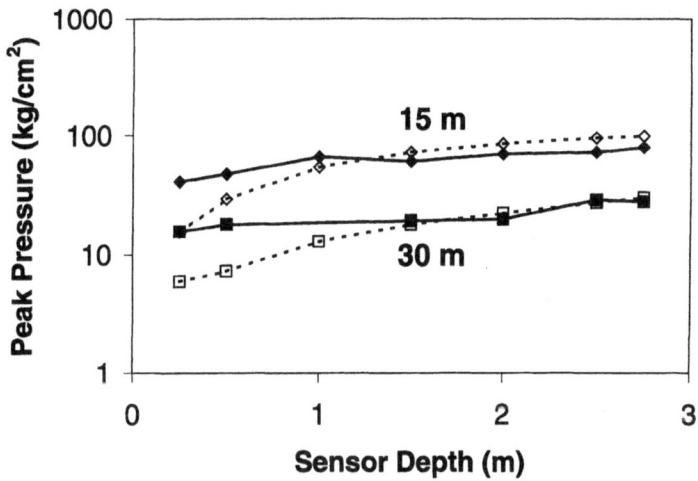

Figure 5

Observed and modeled peak pressures for a half-immersed charge (explosion depth 0 m), water level 3 m. Measured – solid lines and filled symbols, modeled – dashed lines and empty symbols. Horizontal distance to sensor, as indicated – 15 m (diamonds) and 30 m (squares).

predicted peak pressures significantly underestimate the observations at both 15 m and 30 m.

Of the peak pressure measurements available from all 100-kg TNT explosion tests, we further show observed and predicted values for 12 (out of 29) explosions conducted in the full reservoir (water level 3 m). The effect of explosion depth is particularly important and is depicted in Figure 6 for two fixed sensor depths (shallow depth, 0.25 m, and intermediate depth, 1.5 m). It is evident that the best matches are observed at intermediate gauge depths. For smaller horizontal distances (e.g., 7.5 m), the observed peak pressures near the surface and the bottom appear more symmetric than the modeled ones, since the real bottom was considerably more similar to the air above the water surface than the modeling permitted. Also, the modeling likely includes more nonregular surface reflection, most apparent at the shallower sensor depths (≤1 m). Neither factor, however, is too strong at intermediate sensor depths, so that some of the matches between measured and modeled peak pressures are remarkably good (e.g., sensor depth 1.5 m and horizontal distance 15 m).

4.2. Underwater Nuclear Explosion in the Bay of Chernaya (Novaya Zemlya)

An attempt was made to use REFMS to model the seven peak-pressure measurements depicted in Figure 4. The reported yield and depth of this explosion were 10 kt and 30 m, respectively (*USSR Nuclear Weapons Tests and Peaceful Nuclear Explosions 1949 through 1990*, 1996). The test was conducted in shallow water, with the bottom not deeper than 60 m. No other bottom specifications or

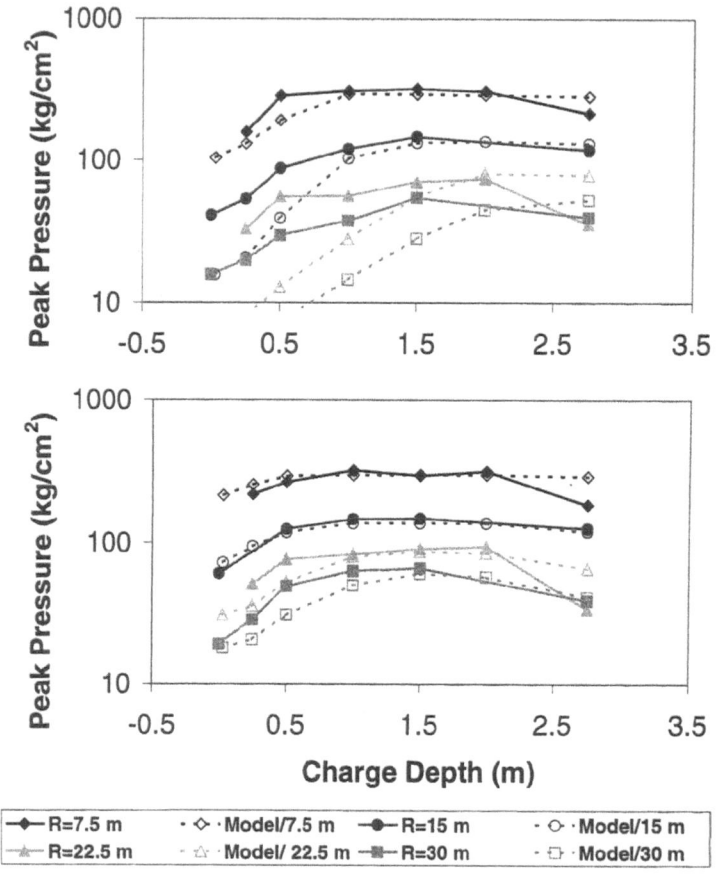

Figure 6

Effect of charge depth on peak pressure. Top – sensor depth 0.25 m; bottom – sensor depth 1.5 m. Measured peak pressures – bold lines and filled symbols; modeled peak pressures – dashed lines and empty symbols. From top to bottom – horizontal ranges 7.5 m (diamonds), 15 m (circles), 22.5 m (triangles), and 30 m (squares), as indicated in legend.

sound velocity profile in the water are available at this time. Thus, using a constant sound velocity in water (1500 m/s), various reasonable bottom velocities and densities were tried, as well as smaller yields and variable explosion depths. Peak pressures were matched quite well, except for the drop at sensor depth of 40 m. This pressure drop could reflect a shadowing effect due to the unknown sound-velocity profile, but could be also due to a measurement error. Despite all uncertainties, this limited modeling revealed that the yield was likely quite smaller than the reported 10 kt, and probably not exceeding 5 kt. Unlike the yield estimate, the explosion depth predicted by the REFMS modeling (between 30 m and 40 m) is in good agreement with the reported depth. The lower yield estimate is further confirmed by the similarity between the peak-pressure measurements at 235 m from this nuclear explosion and the 100-kg TNT explosion at a distance of 7.5 m (around 300 bars in both cases regardless of sensor depth). Thus a scaling factor (COLE, 1948) of about $235/7.5 = 31.33$ may be applicable, which leads to $100 \text{ kg} \times 31.33^3 = 3.08$ kt TNT-equivalent. Accounting for the difference between HE and nuclear sources, the yield estimate for the 1957 nuclear explosion becomes $3.08/0.667 = 4.62$ kt. The factor of 0.667 had been previously determined using a similitude equation for the peak pressure from an underwater TNT shockwave (COLE, 1948); it simply means that a 1-kt nuclear yield is equivalent to 0.667 kt of TNT, with respect to the underwater shockwave.

4.3. Simulated 10-kt Underwater Nuclear Explosions

We would like to use the data and the available modeling capabilities to predict the variations in the hydroacoustic signals that would be observed at the IMS stations. These might differ from peak-pressure measurements in the strong-shock regime in the near field. In the absence of any more detailed observations from real underwater nuclear explosions, modeling results were obtained here for the realistic case of a 10-kt yield detonated in an ocean with a velocity profile modeled with 29 different layers, as shown in Figure 7. This profile (STEVENS et al., 2001) is based on measurements in the Pacific Ocean between Hawaii and the California coast and represents a typical sound velocity profile. It features a minimum velocity at a depth of 700 m, consistent with the SOFAR channel in the ocean. An additional layer was used to model the bottom, with sound velocity 6100 m/s, density 2750 kg/m^3, and thickness 10,000 m.

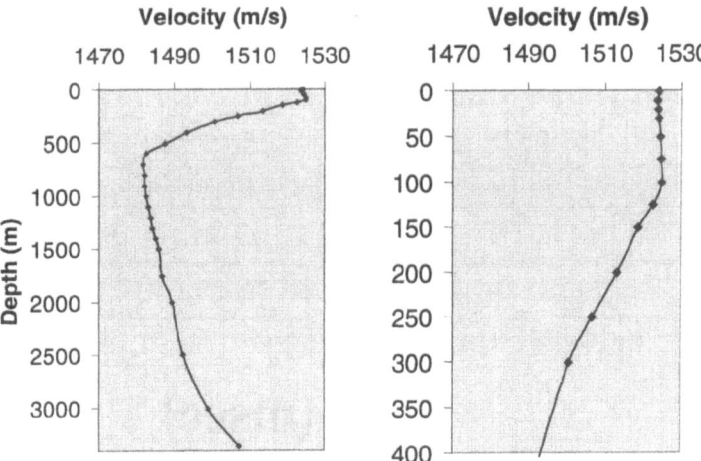

Figure 7

Sound velocity profile used in the REFMS simulation of pressure-time histories from 10-kt nuclear explosions.

The pressure-time histories were calculated at a hypothetical depth of 700 m (i.e., in the SOFAR channel) and a 10-km distance from the explosion. This distance is outside the nonlinear region and is expected to be a reasonable approximation to the signal that would propagate further with low attenuation to the IMS stations. Figure 8 shows examples of pressure-time histories versus depth of the simulated 10-kt nuclear explosions.

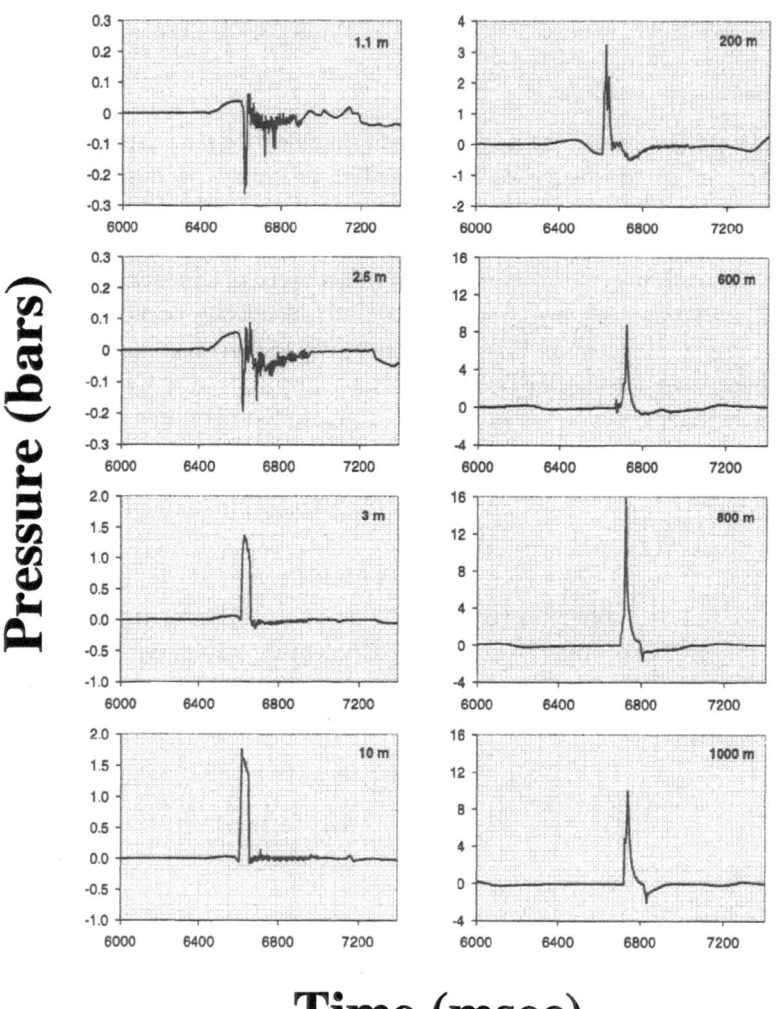

Figure 8
Simulated pressure-time histories recorded at 10-km distance and 700-m depth (i.e., in the SOFAR channel) from 10-kt TNT explosions conducted at various depths as indicated in the top right corners of the plots. The sound-velocity profile from Figure 7 is used in the simulation.

The peak pressures decrease by more than an order of a magnitude for explosion depths shallower than 3 m. It is uncertain how realistic this large drop for surface explosions is. At least part of the predicted strong effect is due to the modeled source being nuclear rather than a HE source. In the depth range 3 m to 800 m, peak pressures increase more than tenfold, after which they decrease at a slower rate. Thus the predicted features are consistent with channeling expected in the SOFAR channel.

5. Discussion

The relevance of the small-scale Russian experiments to larger explosions can be determined on the basis of commonly used scaling relationships, relating distances and times with the cube root of the yield (COLE, 1948). That is, the peak-pressure measurements at a distance of 30 m from a 100-kg TNT charge detonated at 1 m depth in a 3-m deep reservoir should be comparable, for example, with measurements (1) at a distance of 646 m from a 1-kt TNT explosion at 22 m depth above a 65 m deep bottom, (2) at a distance of 1400 m from a 10-kt TNT explosion at 46-m depth above a 140-m deep bottom, or (3) at a distance of 30 km from a 100-kt TNT explosion at 1000-m depth above a 3000-m deep bottom. Factors of $21.5 = (1 \text{ kt}/100 \text{ kg})^{1/3}$, $46.4 = (10 \text{ kt}/100 \text{ kg})^{1/3}$, and $1000 = (100 \text{ kt}/100 \text{ kg})^{1/3}$, respectively, are used in this comparison. The scaling in (3) above is the only one somewhat approximating the ocean environment; however, the sound velocity in water was constant in the reservoir, unlike the real sound-velocity profiles in the ocean featuring velocity gradients.

We would further like to compare the Russian observations, the REFMS modeling results obtained here, and the LLNL modeling results reported by CLARKE et al. (1995). Direct comparison is not possible, because the Russian measurements were taken at considerably smaller scaled distances ($160 \text{ m/kt}^{1/3}$ to $646 \text{ m/kt}^{1/3}$) than the distance in the LLNL modeling ($10,000 \text{ m/kt}^{1/3}$). This was dictated by the limited dimensions of the reservoir. The difference between homogeneous and refractive water is another complication. However, some general conclusions can be still drawn.

CLARKE et al. (1995) assumed a 5000-m deep ocean in their calculations of the coupling factor, using the total wave energies at 10-km distance from 1-kt nuclear explosions with depths varying from 0 m to 1000 m. A mid-latitude sound-velocity profile was used without incorporating bottom interactions. To match their modeling, we performed here a REFMS modeling of a 1-kt explosion using the sound-velocity profile shown earlier in Figure 7. Since the REFMS code allows for different explosion sources to be modeled, both a 1-kt nuclear and a 1-kt TNT source were modeled. This was done to address the possible difference in the way coupling of energy to water changes with depth for HE and nuclear explosions. This change is expected to be smaller for the HE explosions than for the nuclear underwater sources. The REFMS predictions in both cases (HE and nuclear) were calculated for explosions with varying charge depth, at 10-km distance and 700-m depth. The explosion depths in these REFMS simulations matched those used by CLARKE *et al.* (1995).

Figure 9 shows the peak-pressure predictions from the LLNL (CLARKE *et al.*, 1995) and REFMS modeling of 1-kt explosions, together with the previously performed modeling of 10 kt and the most representative Russian observations in this context. The Russian data appear in a separate cluster in Figure 9 as they represent measurements closer to the source. The REFMS predictions for a 1-kt nuclear source match the LLNL modeling for surface (0 m) and deep explosions (1000 m) quite well; about 95% decrease with depth is found in both cases. However, the REFMS change is much steeper for subsurface explosion depths. That is, in the REFMS modeling of the nuclear source, full coupling is approached considerably faster than in the LLNL modeling. Otherwise, the 1-kt TNT REFMS predictions remain above the predictions for the nuclear source by a factor of 0.667 (COLE, 1948) as expected; for explosion depth of 1000 m, the HE peak pressure is 2.77 bars and the

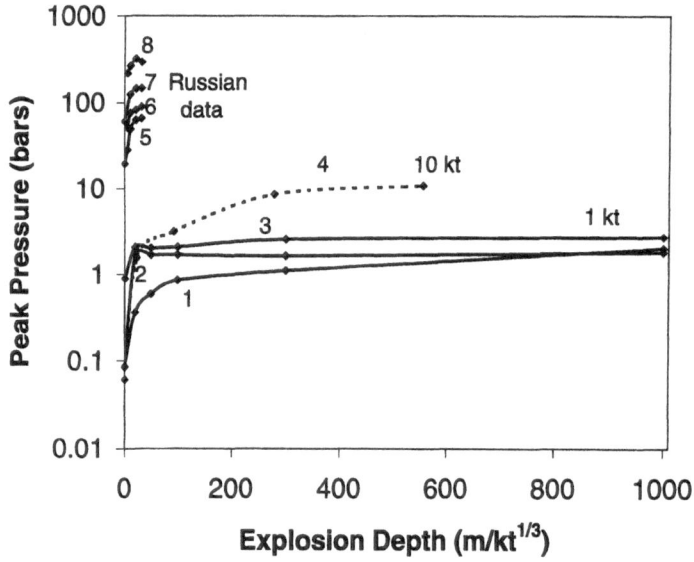

Figure 9

Summary of observed and simulated peak pressures. 1 – LLNL simulation of a 1-kt nuclear source, scaled distance 10,000 m/kt$^{1/3}$, mid-latitude sound-velocity profile. 2 and 3 – REFMS simulations, scaled distance 10,000 m/kt$^{1/3}$, sound-velocity profile shown in Figure 7: 2 – 1-kt nuclear source; 3 – 1-kt TNT source. 4 – 10-kt nuclear source, scaled distance 4,640 m/kt$^{1/3}$, sound-velocity profile shown in Figure 7. 5 to 8 – largest observed peak pressures (at sensor depth of 1.5 m) from the Russian experiments, explosion depths ≤32 m/kt$^{1/3}$ (1.5 m), at the following scaled distances: 5 – 646 m/kt$^{1/3}$ (30 m); 6 – 485 m/kt$^{1/3}$ (22.5 m); 7 – 323 m/kt$^{1/3}$ (15 m); 8 – 162 m/kt$^{1/3}$ (7.5 m).

nuclear pressure is 1.85 bars. The latter is rather close to the LLNL's 2 bars at the same depth.

The REFMS predicted change in peak pressure is remarkably smaller for the HE source (67.5%) than for the nuclear source (95%), when explosion depths of 0 m and 1000 m are compared. In addition, the change in the HE case is more gradual. In fact, the rate of change in coupling for shallow explosion depths suggested by the LLNL modeling is closer to the REFMS curve for the 1-kt TNT than to the REFMS curve for the nuclear source. Although our sound-velocity profile may be somewhat different from the profile used by CLARKE *et al.* (1995), the discrepancies are likely due to differences in the modeling than to differences in the water profiles.

Figure 10 is similar to Figure 9, but focuses on the shallow explosion depths. It shows together the 1-kt predictions, the Russian observations in the shallow reservoir at a sensor depth of 1.5 m, and averaged data from the 1957 underwater nuclear explosion in Novaya Zemlya. The scaled distance for the latter is between 160 and 130 $m/kt^{1/3}$, if a yield between 3 kt and 6 kt is assumed, respectively. The rates of change in the small-scale observations agree very well with the rates in the REFMS predictions of the HE source and some of the estimates in the LLNL modeling. As an example, there is 58.6% (at a distance of 15 m) to 71% (at 30 m) decrease in

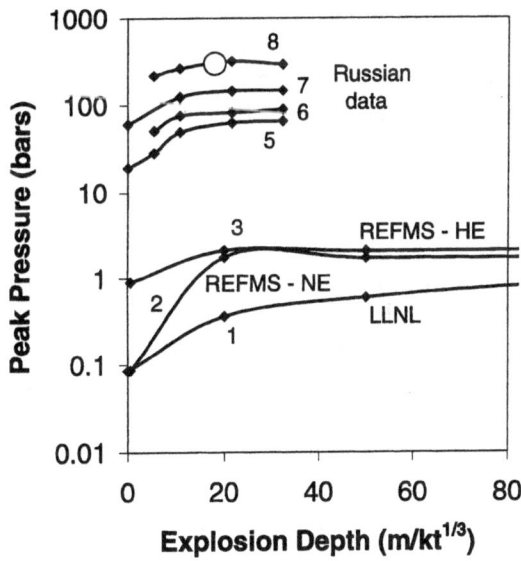

Figure 10
Russian observations and 1-kt simulations, with focus on small explosion depths. Numbers denoting curves – same as in Figure 9. Circle shows the average peak pressure (300 bars) measured at 235-m distance from the 1957 underwater nuclear explosion. A yield in the range 3 kt to 6 kt translates into a scaled distance of 160 $m/kt^{1/3}$ to 130 $m/kt^{1/3}$ (see text for details). This fits well the scaled range of 162 $m/kt^{1/3}$ (real distance 7.5 m) for curve 8 of the small-scale observations, on which the circle falls.

observed peak pressures when explosions on the surface are compared with explosions at mid-pool depths (e.g., 1.5 m), for which the largest peak pressures were measured in the Russian experiments. This matches well the 67.5% decrease predicted by REFMS for the 1-kt TNT explosion when an explosion on the surface is compared with an explosion at 1000-m depth, and is predictably smaller than the 95% decrease for the nuclear sources in both the REFMS and the LLNL modeling. The decrease in peak pressure is the largest for the half-immersed charges in the Russian data and becomes smaller as explosions are detonated at larger depths. For explosion depths 0.25 m, this decrease is from 32% to 62.5% as distance increases, while for explosions at depths 0.5 m, the observed decrease in the peak pressures is from 15% to 25% with increasing range. These data also show that for any fixed shallow explosion depth, a larger decrease is observed in peak pressures measured at increasing distances from the source, but the details of this relationship are not known at present.

The REFMS modeling performed in this work made it possible to make detailed predictions directly relevant to the experimental setup of the Russian tests. The agreement between predictions and measurements provides a further validation of the REFMS code. Given that the modeling of these small-scale hydroacoustic data was satisfactory, especially for mid-reservoir depths of explosions and sensors, it is possible to attempt to calculate the peak pressures that would have been observed at larger distances, given that the reservoir would have also been deeper. The necessary scaled distance here is 465 m (scaled down from 10 km by a factor of 21.5). For that distance and a sensor depth of 1.5 m in a hypothetical reservoir 100-m deep, with everything else the same as in the original Russian experiments, the simulated peak pressures from 100-kg TNT explosions with depths varying from 2.5 m to the surface change from about 2.8 bars to 0.94 bars, i.e., the decrease is about 67%, same as the observed one. Peak pressures calculated for sensor depths of 50 m (mid-depth of the hypothetical reservoir) stay at about 2.8 bars for explosion depths between 5 m and 50 m. This estimate is rather close to the peak pressure of 2 bars for the reference 1-kt explosion at 1000 m depth obtained by CLARKE *et al.* (1995).

Good agreement is also achieved if we take the peak pressures measured in the Russian experiments and attempt to extrapolate them to the hypothetical distance of 465 m, assuming the simplest decay with distance, that is $\sim 1/R^{1.13}$ (COLE, 1948) where R is the horizontal distance from the explosion. The measurements in the middle of the reservoir (i.e., least affected by the boundary conditions), such as for explosion and sensor depths 1.5 m, vary roughly between 300 and 60 bars for ranges 7.5 m to 30 m (see Figs. 2 and 6). The decay of these peak-pressure measurements with distance can indeed be described as approximately proportional to $1/R^{1.13}$, translating into a peak pressure of around 2.8 bars at 465 m distance; i.e., the same estimate as above. This is reassuring, as the $1/R^{1.13}$-proportionality is known to be valid only for unbounded homogeneous water, and our case, although homogeneous, is anything but unbounded.

CLARKE et al. (1995) estimated the total acoustic energy at 10-km range from a fully-coupled reference explosion with depth 1000 m to be 31.3 t. It gradually decreases with explosion depth, so that for explosions at depths 20 m and 0 m (on the surface), the total wave energy is 2.1 t and 0.174 t, respectively. Using these estimates, we can deduce that the energy coupling ratios are about 1:14 for a 20-m explosion depth and 1:180 for a surface explosion. The ratios in terms of peak pressures for the same explosion depths are 1:6 and 1:24, respectively. That is, 13 times change in the coupling energy ratio translates into 4 times change in peak-pressure ratio. On the basis of the energy coupling ratios (CLARKE et al., 1995), we estimated that the total wave energy from the 1-kt shallow explosions at 20 m and 0 m depths is equivalent to that of about 71.4 t and 5.6 t HE, respectively, detonated at 1000 m depth. In view of this, since even kilogram-size HE explosions can be detected under favorable conditions, the IMS hydroacoustic network should not miss 1-kt explosions detonated at any depth in the ocean, including on the surface.

Given the inherent limitations of comparing near-field measurements with far-field simulations, and the additional shortcomings of comparing homogeneous, but severely bounded water, with sound-velocity profiles characteristic for the real ocean, we can assume on the basis of the above discussion that a reasonable agreement exists between the Russian observations and both the REFMS and the LLNL predictions. Thus the reservoir experiments and the REFMS modeling performed here can be considered as a further confirmation that a 1-kt TNT explosion detonated at any depth in the ocean will be detected by the existing IMS network.

Acknowledgements

Bob Britt from SAIC is thanked for discussions and suggestions related to the applications of the REFMS code. Chuck Wilson from Maxwell Technologies is acknowledged for helping identify the relevant literature and available DNA tools. Y.V. Poklad from IDG is thanked for assisting with providing the hydroacoustic data. This work was supported by Defense Threat Reduction Agency contract DSWA01-97-C-0166.

REFERENCES

BRITT, J. R. Application of Cagniard theory to reflection and refraction of explosion shockwaves in soil and water. In Proceedings II 9-th Int. Symposium on Military Applications of Blast Simulations, Oxford, England, 23–27 September 1985.

BRITT, J. R., EUBANKS, R. J., and LUMSDEN, M. G., Underwater Shockwave Reflection and Refraction in Deep and Shallow Water. Volume I – A User's Manual for the REFMS Code (Version 4.0), SAIC (Science Application International Corporation), DNA-TR-91-15-VI, June 1991 (U).

BROCKHURST, R. R., BRUCE, J., and ARONS, A. B. (1961), *Refraction of Underwater Explosion Shockwaves by a Strong Velocity Gradient*, J. Acoust. Soc. Am. *33*, 452–456.

CAGNIARD, L., FLINN, E. A., and DIX, C. H., *Reflection and Refraction of Progressive Seismic Waves* (McGraw-Hill, New York, 1962).

CLARKE, D. B., WHITE, J. W., and HARRIS, D. B. (1995), *Hydroacoustic Coupling Calculations for Underwater and Near-surface Explosions*, Lawrence Livermore National Laboratory, UCRL-ID-122098, 1995.

COLE, R. H., *Underwater Explosions* (Princeton University Press, Princeton, New Jersey, 1948).

KOROBEINIKOV, V. P., and KHRISTOFOROV, B. D. (1976), *Underwater Explosions*, Hydromechanics *9*, 54–119 (in Russian).

KOZACHENKO, L. S., and KHRISTOFOROV, B. D. (1970), *Shockwaves in a Shallow-water Reservoir*, J. Appl. Mech. Tech. Phys. *4*, 166–171 (in Russian).

Research Required to Support Comprehensive Nuclear Test Ban Treaty Monitoring, National Resource Council (NRC), National Academy Press, 1997.

STEPHENS, T., and KELLY, C. S. (1995), *DNA Updates Computational Aids*, Science and Technology Digest, 29–30.

STEVENS, J. L., BAKER, G. E., COOK, R. W., D'SPAIN, G., BERGER, L. P., and DAY, S. M. (2001), *Empirical and Numerical Modeling of T-phase Propagation from Ocean to Land*, Pure appl. geophys. (this volume), 2001.

WALKER, D. A., MCCREERY, C. S., and HIYOSHI, Y. (1992), *T-phase Spectra, Seismic Moments and Tsunamigenesis*, Bull. Seismol. Soc. Am. *82*, 1275–1305.

USSR Nuclear Weapons Tests and Peaceful Nuclear Explosions 1949 through 1990, Ministry of the Russian Federation for Atomic Energy, ISBN 5-85165-062-1, 63 pp., 1996.

(Received June 30, 1999, revised January 14, 2000, accepted January 21, 2000)

 To access this journal online:
http://www.birkhauser.ch